Praise for
ABA Consumer Guide to Obtaining a Patent

"DIY patenting is a disaster, but money-is-no-object patenting is not much better. Smart entrepreneurs (and smart patent attorneys) understand that the more that inventors and entrepreneurs know about the patent system, the better the outcomes for everyone. Rich Goldstein has deep experience helping inventors get the patents they need, and avoiding the ones they do not. If you want to know about patenting, you should read his book. Clear, thorough, but jargon free, it will help you understand the patent system, work effectively with your patent attorney to get good patents, and avoid weak and excessively expensive ones."

—Neil Milton, **author of** *Intellectual Property Law for Dummies*

"Hey inventors, forget do-it-yourself patent applications and kits! Been there, done that. But being informed about the what, where, when, and why of the patent application process is key. What I like best about the *ABA Consumer Guide to Obtaining a Patent* is that this is not a "how-to" book. Rather, it lays out what anyone serious about their invention should know in order to make the best possible decisions about how to proceed. Educated inventors make the best clients for patent lawyers and this book teaches you enough to be a great client, which helps keep fees down. Buying this book might be one of the most successful investments you've ever made."

—Kevin Lee, **founder of Didit and CEO of We-Care.com**

"The *ABA Consumer Guide to Obtaining a Patent* is a fun and stimulating read. The author's years of listening to inventors resonates throughout this book. Our country was built on innovation and this book will help innovators make good decisions. Whether an entrepreneur, a new inventor, or a seasoned pro, this book is a 'must have.' A down to earth resource that can help those who may not know where to start as well as those who strategize."

—Donna P. Suchy, **chair, American Bar Association Section of Intellectual Property Law**

"Rich Goldstein and the ABA provide a much-needed and lucid patent guide for entrepreneurs and innovators, translating often complex and jargon-filled patent arcana into clear and digestible concepts. The *ABA Consumer Guide to Obtaining a Patent* is a great handbook for laypeople to understand the requirements, process, *and* also the benefits and limitations of patents for protecting and exploiting innovation."

—Wayne Sobon, **past president, American Intellectual Property Law Association (AIPLA); owner of Wayne Sobon Consulting**

"Rich has done an excellent job developing an easy-to-read guide to patenting for today's entrepreneurs and innovators. Unlike patenting handbooks, this guide does not confine itself to *how to get a patent*, but also expends significant energy on *why get a patent*. The discussions relating to the business of intellectual property, e.g., licensing, attracting investors, monetization, and so on, are critical to understanding the place patents occupy in supporting economic development, innovation, and entrepreneurship. I highly recommend this guide to patent novices and experts alike."

—Phyllis T. Turner-Brim, Vice President and Chief IP Counsel, Intellectual Ventures

"A great resource and guide for everyone interested in patents, in learning how the patent system works, and in practical, business-oriented answers to the many, many questions that come up from the start to the finish of protecting an invention"

—Philip T. Petti, Chief Intellectual Property Counsel, USG Corporation

"Written in a conversational style and using plain language, Rich Goldstein's *Consumer Guide* makes it easy to understand the complex procedural and legal principles that govern U.S. patent law."

—Alan Kasper, coauthor of *Patents After the AIA: Evolving Law and Practice*

"From my experience in patent valuation, there is a great need for a 'go-to' resource regarding both the basics of obtaining legal protection and the various methods for monetizing these assets. Rich Goldstein's book is the only one I have seen that covers these vital topics in a clear and easy-to-read format."

—Roy D'Souza, Managing Director, Ocean Tomo LLC

"For entrepreneurs of all kinds, nothing is more confusingly murky, painfully stressful, and yet *critical* than protecting their opportunity through patenting. Richard cuts through the fog and makes it clear and simple for a business owner to move forward with their best ideas confidently. Any innovator who is serious about bringing an idea to market needs this book on their shelf!"

—Peter Shallard, "The Shrink for Entrepreneurs," business psychologist and consultant

"The question of 'do-it-yourself' versus hiring a lawyer is ever-present for entrepreneurs, as we're often on a tight budget and inclined to take matters into our own hands; we're often skeptical of lawyers, who may try to sell us too much lawyering for our specific needs. In this book, Rich Goldstein shows us the perfect way to get the best of both worlds: the security of knowing that we've put our IP on a solid-footing legally, on a budget as close to DIY as possible, with as little unnecessary lawyering as possible. Bravo. I look forward to many more books like this from Goldstein."

—Michael Ellsberg, author of *The Education of Millionaires* and coauthor of
The Last Safe Investment

"A few years ago, I filed ten patent applications for a startup I cofounded. I didn't understand the process and soon we received eight rejections, costing the company a lot of time and money. If only I had this book back then it would have worked out much better! Now, after reading this book I know exactly how to handle patenting my next idea."
—Jian Tam, serial entrepreneur

"As an IP attorney from Mexico, it is rarely quick or easy to learn about the patent systems of other jurisdictions. For anyone wanting a deep understanding of how the U.S. patent system works, this book is a must have."

—Juan Rodrigo Pimentel, Arochi & Lindner, Mexico City

"I often hear from my inventor and entrepreneur clients about the many self-help patent guides they encounter. These books were written years ago, are outdated, and full of conflicting information that can cause issues. Their era is over. Rich Goldstein's book is contemporary, up-to-date, practical, and one that every inventor should have in their arsenal for protecting a concept and then bringing that concept to market. This is accomplished with great detail and honesty, and illustrates the innovation and patent system in a light that many must see. I for one will be recommending it to all my clients and fellow attorneys as the one-stop-information shop for everything patents."

—David Postolski, patent attorney, Gearhart Law LLC

"Got an idea for a product that you'd like to patent? You need this book. A simple to understand guide that clears through all of the clutter so you can know if you need a patent, how to get one and most importantly, how to protect your ideas."

—J.R. Fisher, internet entrepreneur; President of Survivalcavefood.com

"Protecting and growing your business can come down to using the law wisely. Patents are one such example where you *must* understand the process to save yourself time, money, and avoid risking your valuable ideas! Rich's book is a step-by-step, very complete guide to what you need to know about patents, and your intellectual property. Understanding this can not only increase the value of your business—but it may make the difference of whether you stay in business. I highly recommend it!"

—Pam Ragland, CEO of Aiming Higher

CONSUMER GUIDE TO

OBTAINING A PATENT

A PRACTICAL RESOURCE FOR HELPING
ENTREPRENEURS & INNOVATORS
PROTECT THEIR IDEAS

RICH GOLDSTEIN

ILLUSTRATIONS BY THOM WRIGHT

20 19 18 17 16   5 4 3 2 1

ISBN: 978-1-63425-607-0

e-ISBN: 978-1-63425-608-7

Discounts are available for books ordered in bulk. Special consideration is given to state bars, CLE programs, and other bar-related organizations. Inquire at Book Publishing, ABA Publishing, American Bar Association, 321 N. Clark Street, Chicago, Illinois 60654-7598.

www.shopABA.org

Contents

Preface

When I set out to have my house built, I wanted to make sure to do my part to make a home I really loved. I started to research. I looked at lots of books and read many articles. I was looking for examples of great design while keeping an eye out for enormous mistakes and lessons others learned the hard way. I didn't study the details of framing or electrical codes or roofing because I knew I was never going to pick up a hammer. I knew I was going to hire an expert to do that.

The thing about people you will hire to pick up a hammer is that they know the important details about how to get the job done once you tell them what the job is. You tell a framer where you want the wall, and he knows exactly what to do. The thing the framer rarely asks is "why?" But often, knowing why would help you get a better result. This is because there are things the experts know that might change what they do for you, if they knew why you want it that way. For example, if the framer knows you are asking for a wall here because you are considering putting in a second kitchen later, he might suggest framing a double wall to allow enough room for the plumbing. Sometimes when you know a little about what they do, why they do it, and how they think—and they know why you want what you want—that knowledge will go a long way toward helping you get the result you are looking for.

Acknowledgments

What an honor to be the one who gets to write this book on behalf of the American Bar Association! Thanks so much to ABA Publishing and the IP Section Book Board for this amazing opportunity. Kevin Commins: I don't know exactly why you thought that I should be the one to write this consumer guide for the American Bar Association, but I'm grateful that you asked, and plainly none of this would have happened if you didn't!

Thom Wright: thank you for being such a great friend and insightful advisor. As always, your patent illustrations look amazing and add so much to this book.

I appreciate so much all of the help and support generously given by Jennifer Sutter, Phyllis Esposito-Driscoll, and Gregg Brown. Thank you, Sunny Zebe, for your tireless editorial help. Gerry Broard, thank you for your help and friendship. Thank you, Sharra Brockman and James Hanft, for taking the time to review this manuscript and for your helpful suggestions.

In writing this book, I am grateful for the support of my family, friends, and colleagues. To my wife Danielle and my children Brianna, Evan, and Julia—you make it all worthwhile. Of course, my parents, my brother David, and my sister Donna are such a fundamental part of everything I do and whatever I may accomplish. A special thank you to my good friend Steve—I've learned so much from you.

Introduction

Is This Book for You?

If you are an inventor or entrepreneur who wants to understand how to use the patent system to help fulfill your business goals, this book is for you. If you have an idea that's important (*valuable*) to you, to your business, or to the start-up venture you have in mind, you are *not* going to try to patent it yourself. Smart entrepreneurs use patent attorneys. Really smart entrepreneurs use their own time effectively, learning just what they need to know to understand the process, so they can use their patent attorney's time effectively and work the system to their advantage.

Entrepreneurs with valuable ideas hire professionals to help them. But that doesn't mean they always spend their money efficiently or wisely. Before you do invest in a patent, it pays to invest a little time and learn how the patent system works. Reading this book will help you understand (1) how to work the patent system to your advantage, and (2) how to work effectively with the patent attorney who will represent you. As a result, you might save yourself a lot of time, money, and aggravation.

You might not know it, but the American Bar Association (ABA) publishes a lot of books! However, they're mostly books written by lawyers to explain to other lawyers the intricate details of a particular area of the law.

This book is different. The "ABA Consumer Guide to Obtaining a Patent" is written for *consumers*—non-lawyers who want a better understanding about how to use the patent system effectively to obtain a patent.

When we speak of consumers, typically we are speaking of people who are in the process of buying, or at least contemplating making a buying decision. In this context, it means people who are considering applying for a patent and who may be considering hiring professionals to help them secure their patent rights. If that's you—a potential consumer of patent services—then I think you'll find this book to be quite helpful.

What Will You Learn by Reading This Book?

Think of the manager at a mid-sized company that hires patent attorneys all the time to protect its important innovations:

- How much does *that guy* know about the patent process?
- What type of experience does he have that helps him make good decisions about *what* to patent and *when*?

That's the benchmark for this book. I want you to know as much as *that guy*, so you can make equally effective and efficient decisions *with confidence*— as if you've done this many times before. If your patent attorney recommended this book to you— or perhaps even gave you a copy—it's probably because your attorney knows that the better you understand the process, the better client you will be. The more you understand the dynamics of the patent system, the more likely you are to make the most of it, and of his or her representation of you.

By reading this book, you will learn:

- *What* you might already be doing *without realizing* that can jeopardize your chances of ever getting a patent
- *How* to determine if you can get a patent on your invention
- *Why* a patent may or may not be important for your particular business goals
- *When* to file a *patent application*, and at what point you should get an attorney involved
- *How to avoid* making the patenting mistakes that even smart entrepreneurs make every day

This book won't teach you how to write a patent application. What it *will* teach you are the most important principles to understand about patent applications, and about the patent system, if you want to protect your patentable ideas effectively and efficiently—and avoid wasting time and money.

Before I went to law school, I thought that it would be about studying and learning all the laws I would need to know in my professional career. When I got there, however, I found out that in law school they don't try to teach you everything a lawyer would need to know to competently practice law. Instead, they train you to *think like a lawyer* and to effectively use various available resources to find out what you need to know, when you need to know it.

Similarly, my intention here is not to teach you all of the detailed rules about patents. Instead, I want to give you a framework and an understanding of the resources you will need so that you can begin to think like a patent lawyer—or better yet, think like an entrepreneur who is "patent savvy."

Why I Wrote This Book

Over the past 20+ years as a patent lawyer, I've had the amazing opportunity to represent thousands of entrepreneurs, and helped them obtain nearly 2,000 patents. I've counseled more than 10,000 inventors about the patent process one on one and educated at least another 10,000 through my online videos.

Along the way, I've learned what entrepreneurs *need* to know about patents to be successful. But even more important than that, I've learned what entrepreneurs *want* to know about patents so that they can make decisions they are satisfied with, and then turn their attention to starting and growing their companies. I am proud to have the opportunity, through this book, to share my experience and insights with you.

What to Expect in Reading This Book

Most of the book is written like a conversation, just between you and me. I recognize that if you're reading this, it's probably not just an academic exercise for you; you are actually considering getting a patent. Because of this, I feel it appropriate to be a bit informal and to speak directly to you.

You'll find that I tend to use the words "idea" and "invention" (and sometimes "product") interchangeably. In doing so, I run the risk of offending the purists who say, "You can't patent an idea, you can only patent an invention." Well, they are technically correct. You can't patent, for example, the fanciful idea of *a machine that can fly*, but you could get a patent when you've taken it a step further and conceived of an actual structure that is capable of flying, like an airplane or a helicopter. So yes, you can patent an idea, when it is more than just an abstract notion but is a real-world manifestation. Now, with that out of the way, for the rest of this book, I'll just use "idea" and "invention" interchangeably.

Also, I'm not going to constantly hit you over the head with "you've got to use an attorney."

Moving forward, I'm going to assume that protecting your idea is important to you, and your intention is to do it right. That being said, here's the disclaimer: This book is not intended to be legal advice. It is intended to inform you about the principles of patenting. Legal advice, and appropriate actions to establish and preserve your legal rights, should only be taken upon consulting qualified legal counsel who is fully aware of your facts and circumstances.

Now, let's get started by discussing *why* you might want a patent.

Do You Need a Patent?

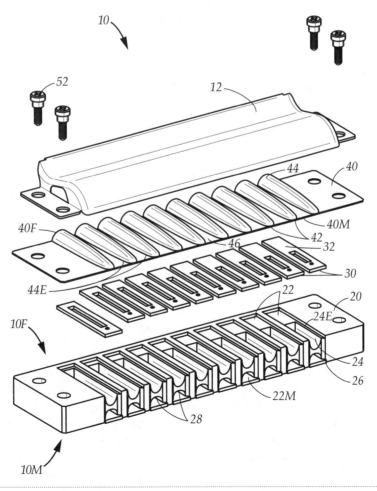

"I want to apply for a patent. How do I get started?" the caller asked me.
"Why do you want to get a patent?" I inquired.
Long silence . . . and then:
"Isn't that what you're supposed to do when you invent something?"

If you do pursue a patent, it's not going to be free! So it pays to first carefully consider and understand the real reasons you would do it.

What Can a Patent Actually Do?

In simple terms: a patent is a grant of rights by the government, for a limited time, that can be used to *stop others* from making, using, or selling your invention.

Here's where we can begin separating myth from reality: having a patent *does not* give you the right to legally make your invention! Said differently, having a patent is not a requirement for you to make or sell your invention!

What a patent can do, however, is give you the right to stop others from making, using, or selling it without your permission. Now that this is clear, we can take a look at whether the ability of a patent to stop others from making, using, or selling your invention is important enough to *your goals* to justify your decision to patent it.

Reasons to Patent Your Idea

Right or wrong, here are some common reasons people and companies seek patent protection for an idea:

- The fear that others will steal it
- To generate licensing revenue (royalties)
- To prevent or reduce competition
- To maintain or acquire market share
- To enhance company valuation
- Because investors want it
- For business credibility or marketing
- For personal credibility or vanity
- For the experience
- Because someone told you that you should
- To avoid infringing someone else's patent

Knowing *your* reason can help you decide whether it is worth the time, money, and effort to seek a patent. It can also help you and your attorney pick the right strategy in the beginning of the process, as well as make appropriate decisions along the way.

Let's consider the merits of each of these common reasons for seeking a patent. As we take a closer look, some of them are more compelling than others. When people talk about diet or exercise, they generally talk about "healthy choices" and "unhealthy choices" (rather than "good" and "bad"). So let's talk about these reasons in that way. As an entrepreneur, for your business, some of

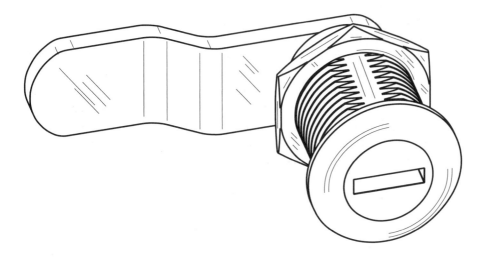

these reasons are "healthier" than others. Let's just say a few of them are somewhat unhealthy—and at least one of them is dead wrong!

The Fear That Others Will Steal It

Immediately after realizing "I have an idea," nearly every inventor thinks, "But I don't want anyone to steal it!" With the notion that patenting is the way to stop other people from stealing ideas, many inventors quickly decide: *I need a patent.*

While often there are practical reasons why someone stealing your invention would be a bad thing, it's usually the actual fear—that is, the emotional/ psychological part—that's the most powerful and the most compelling. Since this is probably the main motivation for most individuals seeking patents, let's get a little psychological here for a moment to better understand this dynamic:

> Gripped by the fear of losing out, inventors generally never examine how—or whether—having someone else pursue their invention without their permission (i.e., "steal their invention") would actually have a real impact on their goals.

Most of the remaining reasons discussed below describe real scenarios where having someone else "steal it" would actually undermine some goal or result that you are attempting to accomplish. Before we consider those other reasons, we want first to notice *just how powerful the fear itself is* so that we can understand its grip over us. Then we can put the fear aside and take a measured look at the other, more logical reasons that getting a patent would help you reach your goals.

To get an immediate sense of what this fear can feel like, imagine a future where you created something with the potential to take you places—to bring you wealth, fame, and prominence in your field. Then imagine that instead, in this future, you never get to enjoy any of it because someone else steals your idea and reaps the benefits of your creation. Thinking about this scenario, do you feel a little sick to your stomach? Well, this is exactly what most inventors imagine, which leads them to hastily decide, *I NEED a patent!*

Ultimately, there's nothing wrong with seeking a patent to avoid the devastating experience of knowing your idea was stolen. It's helpful, however, to know whether that's your *primary* motivation. Knowing your motivations gives you more choice in the matter—and then at least you'll be clear about what you're paying for! Also, understanding your motivations allows you to make choices that are more conscious, more deliberate, and a better fit for your current circumstances.

For example, say that sometime after you filed your patent application, you come to the realization that "fear of theft" was the thing that originally drove you to seek protection. Then, at some later point in the process, perhaps you determine that it's unlikely that anyone will steal your idea, or that if they did, it wouldn't actually interfere with your plans. Knowing now that there isn't a real *economic* reason to keep pursuing the patent project, when the expenses start adding up, you might make a conscious choice to drop it—and stop spending money unnecessarily on a patent you realize you don't actually need.

To Generate Licensing Revenue (Royalties)

A frequent strategy for profiting from an invention is to find a company that will manufacture, market, and distribute the product, and pay the inventor a *royalty*. This highly sought-after arrangement is called "licensing." The way it works is, you sign a contract giving another person or company the right to use your idea in exchange for royalty payments to you that continue for the term of the contract. But to get a licensing deal, you must first establish that you hold the rights to the idea—because those rights are the thing that the company would license from you. Most typically, it's *patent rights* that are the subject of the licensing agreement.

It's not just inventors who follow this strategy. Many major corporations have licensing as a central part of the company's business model. Most large companies that research and develop new technologies patent their ideas, without ever intending to actually produce products from many of them. Instead, these companies actively seek out "licensees": other companies that would like to use

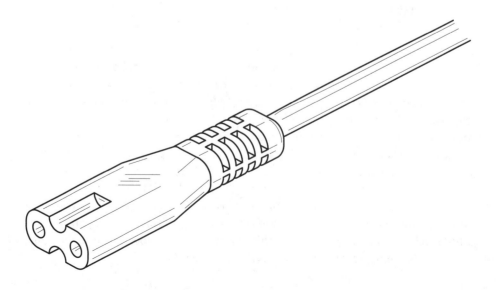

the technology in exchange for a royalty. Following this strategy, some companies have licensing portfolios containing thousands of patents, and they collect billions in licensing revenues.

If your prime reason for seeking a patent is to license it, there are a few things you should consider. First, *you* must be the one to find a company that wants it and is willing to pay you for your idea. In particular, you will need to find *the right person at the right company*—a person who appreciates the value of what you have created and is in a position to manufacture, market, and distribute it. Finding this person will require that you knock on some doors. A common misconception is that there are agents willing to make this connection for you. In reality, it's something you need to do yourself. One truth you should realize as an inventor is that no one will be as passionate about your invention as you are! Before pursuing a patent with the sole intention that you will license it, ask yourself: are you prepared to spend time knocking on doors to find someone willing to license it?

To Prevent or Reduce Competition

As an entrepreneur launching a new product, you want to sell as many units of your product as you can. Clearly, your efforts can be slowed by others competing for the same customers. Preventing or reducing that competition could make selling your product a lot easier.

Talk to anyone who has a company that's been in business for a few years. Ask these business owners: if they could go back a few years and do whatever they possibly could have done to stop the competition they're experiencing now, would they do it? Yes, of course they would! This is because they understand that, now that they are farther down the road, it costs a whole lot more to thwart the competition. Once the competition is breathing down their neck, these business owners constantly have to invest in developing new and better products, and spend more on marketing, to stay competitive.

Often, experienced entrepreneurs launching their products will file patent applications even when the odds of getting a patent are less than 50 percent. They know that for just the *chance* that they will get a key patent, whatever they spend now will be worth it a few years down the road, when they are competing head to head with other companies.

The key to this approach is being able to discern whether you have a *special solution* that could prevent or reduce competition. You may think your invention provides a "special" solution. But the test of whether it's truly special is whether it's something that consumers of products like yours *really* want or need. If your invention does provide the solution that is most wanted, needed, and requested by consumers, your competition will want to do exactly what you're doing! When they can't do what the consumer wants them to do—or include the product feature that the consumer wants them to include—because you hold the patent, you have effectively prevented or reduced competition.

One of the great benefits of even just applying for a patent is the uncertainty that it creates for your competition. When it's not clear whether they can safely copy your product—because they don't know exactly what's in your patent application—they are less likely to make the risky decision to copy it.

To Maintain or Acquire Market Share

Similar in concept to *preventing or reducing competition*, a patent on an important feature can be your foot in the door in a very competitive market. It can also be the reason why a winning product remains relevant in the marketplace.

Many times, companies innovate to become—or to remain—relevant to consumers. When these innovations are patentable, the company's *patent portfolio* provides a kind of *market share insurance*. To consider the effectiveness of this "insurance," the question to ask is: do the features embraced by the patents in their portfolio make their product compelling to the consumer?

As we discussed earlier, it's important that there's alignment between the features you're patenting and what the market wants. The key is to patent the feature or features *that make a difference* and provide a solution to a widespread problem. Providing such a solution makes a product highly desirable in the marketplace or helps maintain its stature. It's this overlap between marketability and patentability that provides the basis for the most valuable patents—and that can have the greatest impact on market share. If, however, the part or feature that's being patented does not make your product significantly more desirable, others can simply provide their own just-as-good solution to the problem you identified and launch their product to compete easily with yours.

To Enhance Company Valuation

A patent can greatly enhance the value of your company because it represents an aspect of the company that others can't freely duplicate. If you build a company with a certain number of customers, level of revenue, and market share, finance people can readily put a number on the value of your company. In part, the way they arrive at that value is through considering:

> What would it take for someone else to start a company in your field and achieve a similar customer base, revenue level, and market share?

This is a fair way to measure the value of a company—*unless* there are patents or other *intellectual property* (IP) involved!

When there are patents involved in the key business of the company, and the company owns those patents, they are the part of the business that others can't replicate. So the finance people can no longer calculate the value as just what it would take to build up a company *like yours*. If you own the key technology, you become "the only game in town." Suddenly, the value of your company multiplies!

In the figure below we consider two examples considering the value of a company based on a pie chart that shows its market share. Often, the market share of a young company is very small, perhaps just a few percent. But when you consider the value of the company, whether it owns the patents on the key technology makes a huge difference. If the company doesn't own patents, and markets

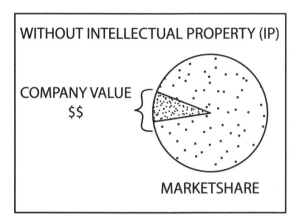

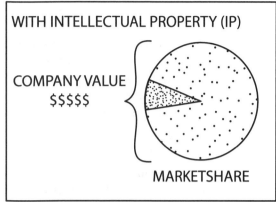

non-proprietary technology—as in the example on the left—the value of the company is based on its small percentage of actual customers; the small piece of the pie. However, if the company does have the patents, then the value of the company can reasonably be based on the other 90+ percent of the market that the company hasn't reached yet. In other words, with patents—as in the example on the right—the value can be based on the *potential* market share—the whole rest of the pie!

On this principle, a portfolio of patents can increase the value of a company, *provided they are good patents and relevant to the core business model.* Moreover, this portfolio can be a great bargaining chip if you decide to sell your company.

Because Investors Want It

On the popular TV reality show *Shark Tank*, individuals pitch billionaire venture capitalists to invest in their fledgling companies. When it's a product that the person is requesting for the "Sharks" to invest in, the first question the Sharks often ask is, "Do you have a patent?"

This is a common question from investors who aren't on TV, too! Potential investors in your idea or company will want to know that you've taken steps to protect your idea. They will be naturally leery of backing you if the primary business asset you have—your idea—is not protected. Before an investor puts money on the line to invest in your company, there has to be something *indispensable* about the opportunity in order to make the investor want to put money in *your* company. It may be you, yourself—and the investor's belief in your ability to capitalize on the opportunity. It also may be the success that your product is already experiencing. But if it's the idea itself that investors find most compelling, they naturally will want to know why putting money into your company is synonymous with putting money into the idea. If the idea is *proprietary* (you own the rights to it), that helps justify why they are investing in your company. If it's not proprietary, however, they will wonder what other options they have to still invest in that idea—*without necessarily investing in your company!*

"Because investors want it" is among the most compelling—and common—reasons that entrepreneurs in start-up ventures seek patents. Consider that whether they are investing in your product idea or just *licensing* it, before anyone hands over their money to you, they will want to know that you actually own what they are buying or buying into.

For Business Credibility or Marketing

To potential customers, knowing your product is patented can increase its stature in the marketplace. Often, product advertisements proudly tout that a product is patented. They do this because many consumers perceive a patent to be a "stamp of approval" from the US government. While it's not really true that a patent indicates government approval or endorsement of a product, having a patent is often perceived as an indication that a product is innovative and exclusive. In some realms, having multiple patents is often seen as lending credibility to the company's position in the field or as an indicator of the company's "reputation for innovation."

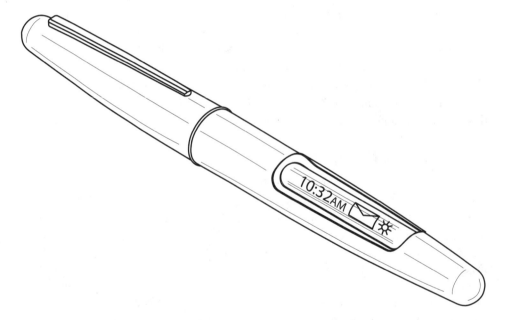

For Personal Credibility or Vanity

Some people seek a patent on an idea just to hang the certificate on the wall—to be able to say, "I have a patent!" There's nothing wrong with that, but keep in mind that patents aren't cheap. It can be pretty expensive wall art. If you're into that, it *is* kind of cool to have one. And it's definitely something to be proud of!

For scientists, engineers, and academics, however, patents are more than just cool wall art—they also add to the individual's credibility as an expert. In certain industries, having several patents listed on your resume or CV will naturally increase your stature in the field.

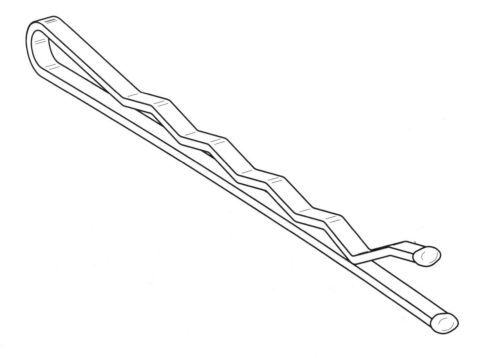

For the Experience

Does pursuing an invention—doing your best to take an idea of your own and get it on the market—sound fun, exciting, or enriching to you? Well, pursuing a patent on an invention for the sheer experience of it might be one of the best reasons to do so! For a serial inventor or fledgling serial inventor, the education you would gain just by going through the patent process may be well worth the investment of time and money. Often, when people are successful with an invention, it wasn't their first invention or their first attempt at launching a product that found success—it was their second, or even their third! However, the first time they went through the process was a valuable experience. Going through the design process, the patenting process, and the marketing process that first time provided them with loads of knowledge and experience that they put to use later, when it was time to pursue the idea that really hit.

Some years ago, I had a client who came into my office and demonstrated an invention that could best be described as a practical joke device—kind of like a whoopee cushion. He had a big smile on his face as he demonstrated it and told me, "I want to patent this!" Now, he didn't have any grand plans about making a fortune with his idea, he just wanted to do it for the fun of it. And so we did. We got him a patent, he had a great time along the way through the whole process, and I had a great time working with him. Isn't it fun to work with people who are having fun? The point is, he went through the patent process just for the experience. Many people would consider that a rather expensive experience! But for him, it was worth it. More power to him!

Because Someone Told You That You Should

Yes, it's true. When many people seeking a patent are asked why, the best reason they can give is *someone told them that they should*. They haven't stopped to consider whether it's something they *actually want*. If you are looking into getting a patent mainly because others are telling you that you should, this is an opportunity to consider whether any of the other reasons mentioned in this chapter reflect something that matters to *you*.

To Avoid Infringing Someone Else's Patent

"I want to patent it, so I know I can make it" is another reason I commonly hear from inventors. Unfortunately, this reasoning stems from a major misconception! In fact, patenting an invention is no guarantee that it won't infringe other patents.

This is a very important concept, so let's talk about it right now! Having a patent *does not* give you the right to make a product. It gives you the right to *stop others* from making, using, or selling the product. There is an important difference here. Consider that someone has a patent on a core part of your idea, but you have made significant improvements on it. Your improvements on the already patented product can result in your getting your own patent. But that doesn't mean you would be allowed to produce your product if doing so requires you to make or use the part that was patented by someone else. In granting you a patent on your improvements, the United States Patent and Trademark Office (USPTO, aka "the Patent Office") does not consider whether you could actually produce your product without infringing any other patent. Thus, getting a patent is no guarantee that you are not *infringing* the patents of others! The dynamic between your patent and theirs will be clearer when we talk about *patent claims* in Chapter 6 and about *infringement* in Chapter 15.

IMPORTANT CONCEPT

There is a difference between patentability and infringement. They are judged by different standards, and have different implications. Patentability is whether you can get a patent. Infringement is whether you are violating someone else's patent. This will be explained in detail in Chapter 15, "Infringement: What to Know and What to Do."

Reasons Not to Patent Your Idea

Your idea may be great, it may meet all the criteria for patentability, and there may be a lot of people willing to pay for it, but that doesn't mean a patent is necessary. It may not even be in your best interest to pursue a patent. Some reasons why a patent might not be worth it include:

- Fast-moving technology/quick obsolescence
- Better use of resources elsewhere

- Market for the product is too small

You may discover other reasons as you get deeper into the patent process.

Fast-Moving Technology/Quick Obsolescence

In some fields, technology moves very quickly. A key innovation—a blockbuster product today—is forgotten tomorrow. It sometimes takes years to get a patent. If your idea truly is something that will be a short-lived fad, a patent might not be worthwhile. By the time the patent issues, the idea might already be irrelevant. But don't be hasty in judging your product to be a fad. Some products seem like they will be a fad initially, but they end up having significant staying power.

Better Use of Resources Elsewhere

Seeking a patent requires a significant investment of resources. For many new ventures, it may take a large chunk of your available funds. At times, there may be a more compelling need or another investment that provides a more immediate opportunity for growth. This is a time to carefully consider the value of IP against the value of the other possible expenditures of capital. However, while skipping the patent might provide short-term cash flow relief, if the technology becomes key to your business, you might later wish you had invested in IP protection.

Sometimes the patent might provide some value to the venture, but the brand names you are establishing are potentially much more valuable. In that case, shifting resources from patenting toward trademarking might make more sense.

It really depends on your individual circumstances. This is a subject you'll want to discuss with a patent attorney, who may be able to help you weigh the pros and cons of going for a patent. And if your thought is to put off the patenting until later, your attorney can advise you whether waiting is even an option.

Market for the Product Is Too Small

An idea may be great and might turn out to be totally loved and embraced—but by only a limited group of people. If, for example, the product is a tool that would only be used by a certain trade or individuals with a certain hobby, you must ask, *how many people actually practice that trade or engage in that hobby?*

Imagine that an inventor's new woodworking tool would cost $50 and would be useful to a total of 2,000 people. That means the total market for the product, assuming 100 percent of the potential customers actually bought one, would be $100,000. It is, of course, still possible for the inventor to have a successful product here, on a small scale—as long as the expenses involved in producing and marketing the product are kept low. But would it really make sense for the inventor to spend possibly tens of thousands of dollars on a patent? Probably not.

And if, after the inventor patented it, someone else copied the idea and grabbed a piece of that $100K market away from the inventor, would it even be worth it to go after the competitor for infringement? Probably not. So, when the total market for the product is very small, a patent might not make sense unless there is some other compelling reason for protection.

Summary

It pays to consider why you are applying for a patent. Knowing your reasons will help you to decide whether applying for a patent is the right decision and will shape your strategy throughout the process. For more information, visit the companion site for this book: www.patent-book.com.

Can You Get a Patent?

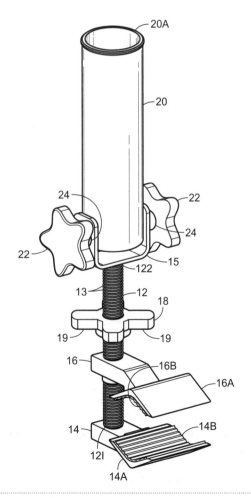

After you've resolved the question of why you want a patent, the next big question is: can you get a patent? Perhaps you've decided that you want to get one, but now we need to figure out if it's possible to patent your invention.

There are many principles that will shape our approach in applying for a patent. We will talk about a lot of those in detail in the chapters that follow. What we want to talk about now, however, are the *threshold* questions. The threshold questions help us decide whether, overall, it is a *go* or a *no-go* for applying for a patent. Before we get too deep into the patent process, then, let's take a look at how your patent application will be judged at the United States Patent Office so that you know whether you should even try to get a patent.

The Questions to Resolve First

There are three main threshold questions you should consider before trying to patent your invention:

    1. Does it have *patentable subject matter*?

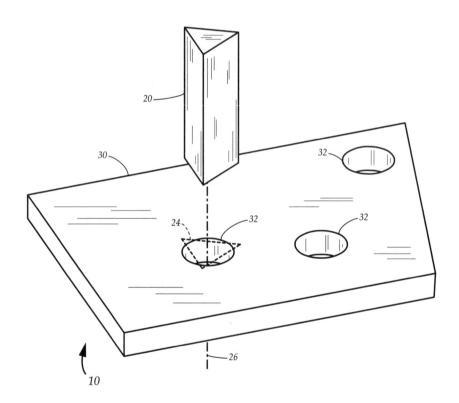

    2. Is it distinctive enough from previous inventions? (Technically, is it *novel* and *non-obvious?*)

    3. Can its operation be explained? (Technically, can you provide an *enabling disclosure?*)

We'll look at the first two in more detail before tackling the third because, for most inventions, the third requirement is just part of the patent attorney's job in preparing your application. But for some inventions, it can actually be a threshold question. After all, if no one knows how your invention would/could work, your attorney won't be able to explain it either. And if that's the case, it cannot be patented.

1. Does Your Invention Have Patentable Subject Matter?

This is the threshold of all threshold questions because, if your invention is not the type of thing that can be patented, then the inquiry stops here! Not all ideas can be patented. In many cases, it's simply because they don't fit one of the categories of what can be patented.

If you are looking to patent an idea for a book you've written, then [*buzzer sound*] Nope. A book can't be patented because it is not the type of thing for which you can get a patent. The same is true for a song. It doesn't matter how original the song is, it's just not the right type of thing for the Patent Office to protect. If you go into the Department of Motor Vehicles and ask for a building permit, they'll tell you that you're in the wrong office. And that's what the Patent Office would tell you: "Sorry, wrong office! We don't do that here. Maybe try the Copyright Office down the street."

So, what types of ideas can be patented? For a *utility patent*, there are four categories of patentable subject matter. To be patentable, the idea must fit into at least one of these categories:

- **Machine**—something with parts that interact. When you think of a machine, you might think of something with gears, something that might

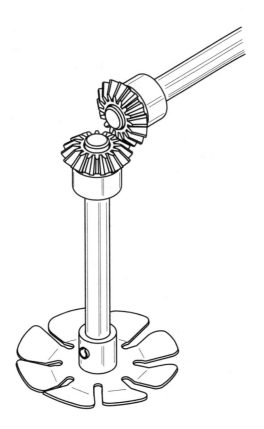

have a motor or an engine, things like that. However, things that qualify as machines include anything with parts that interact with each other. So even an electrical circuit—where nothing actually moves—is considered a machine under patent law and can be eligible for patent protection.

- **Manufacture**—something that doesn't have moving parts but still has a function, such as a paper clip. A paper clip is a single piece of wire that's

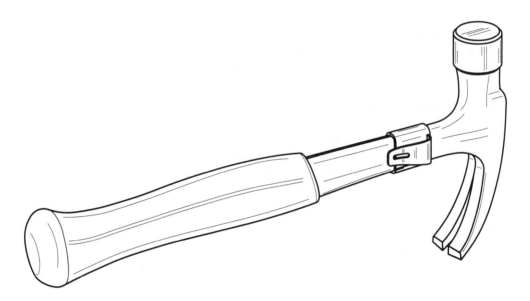

bent in a way that allows it to exert pressure against pages to hold them together. Because it's a functional, physical item, even without moving parts, a paper clip would have patentable subject matter as a manufacture.

- **Composition of matter**—a combination of chemicals. This can be a pharmaceutical, a household cleaning product, a new shampoo, and the like. A composition of matter exists when you take more than one chemical, put them together, and the resulting combination serves a useful purpose.

- **Method/process**—a series of steps for doing something. Traditionally, patented processes include things like a new way of refining steel, where you have a new set of steps that you follow in order to refine the steel. It's

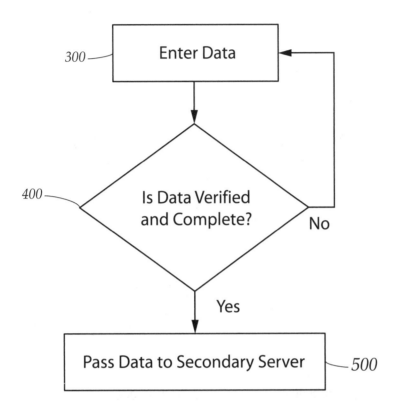

not that the resulting steel is different from what came before; it's that the method or the process *or the way that you do it* is different. Also, over the last couple of decades, software has been protected using the category of process or method. Certain ways of doing business have also been protected because they fit under this category.

If you can't decide whether your invention would be considered a manufacture or a machine, that's okay. Sometimes there is overlap between the categories, and when you apply for a patent, you don't actually need to specify which one it is. As long as the invention fits at least one of the categories, you will satisfy the requirement for patentable subject matter. Things that don't fit any of these categories, however, cannot be patented. By now, it's obvious why a song or a book can't be patented. It doesn't fit into any of the four categories. A book or song typically can be protected by a copyright, however. We'll talk about copyrights and trademarks in Chapter 16.

By the way, another requirement for patentable subject matter is that the invention be useful. But don't worry, just about any use will do. This is rarely an issue and usually only comes up when someone attempts to patent a "perpetual motion" machine, something that clearly could not work as described, or something that otherwise violates the laws of physics.

The Overhand Wheel
(Classic Perpetual Motion Machine)

2. Is Your Invention Distinctive from Previous Inventions?

To be patentable, your invention must be *different enough* from previous inventions. To determine how distinctive your invention is, it has to be compared to the closest previous ideas. The previous ideas that the Patent Office compares your invention to are called *prior art*.

Other ideas that are similar to yours, and that have been made public in certain specific ways defined in the patent laws, are considered prior art. Examples include existing patents in the United States or abroad, certain filed patent applications, publications describing the invention, prior public use of a similar invention, and so on. The rules have gotten very complicated about what may be considered prior art against your patent application. If there is any doubt about whether something would be considered prior art, you should consult a patent attorney. What to know, however, is that things that are publicly available and came out or will come out before the date that you would likely apply for a patent are most likely prior art and should be considered in determining whether your invention would be new enough to get a patent.

Since the Patent Office examiner will search for prior art when he or she reviews your patent application, it pays for you to do your own searching ahead of time to know what you would be up against if you do apply for a patent. A *prior art search* generally focuses on other patents similar to your invention, and it can help

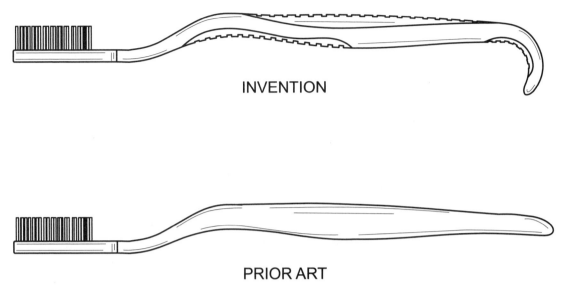

INVENTION

PRIOR ART

Figure 2.1 Determining whether an invention is novel and non-obvious begins by comparing it to the closest prior art

determine what, if any, inventions may create an obstacle to your getting a patent. (See Chapter 3, "Is Your Invention Really New?," for tips on doing a prior art search.)

The existence of one or more pieces of *close prior art* could render your idea unpatentable. However, just because a prior idea is similar to yours doesn't necessarily disqualify your idea from patent protection. It's very rare that any idea is *completely* new, meaning that nothing like it *ever* existed before. Thus, most patents are for *innovations*—improvements on existing ideas. Improvements on existing ideas are potentially patentable. The question becomes, what is the improvement and is it different enough?

Usually when we search, we will find something somewhat similar, often even more similar than the inventor would have ever imagined existed! This is the time to look more carefully at what is *actually distinctive* about your invention. Of course, if the prior art is too close, perhaps not much of your idea will be distinctive and it will not seem patentable over the prior art. When the prior art is just somewhat close, it might not be a game stopper, but it might very well change your strategy. For example, once you learn which parts of your invention are distinctive from the prior art, those distinctive parts are the ones you would probably emphasize in your patent application. In Chapter 6, we'll see how the results of a prior art search can shape how your patent application is written.

Novelty and Non-Obviousness

Let's get a bit more technical now. I've been using the term "distinctiveness" to make the point that your invention must be *different enough* from the prior art to be worthy of a patent. In reality, the Patent Office doesn't consider or use the term "distinctiveness." The examiner evaluates whether an invention is different enough in terms of *novelty* and *non-obviousness*.

In this case, *novelty* refers to whether there is *any* difference between your idea and anything that already exists in the prior art. An invention is novel if *there's nothing exactly like it in existence or documented anywhere.*

It's usually easy to get past the novelty requirement. I mean, if nothing is exactly like yours, it's novel. Once you get past that, however, *non-obviousness* comes into play. It can be a little difficult to grasp the concept of non-obviousness at first, but here is a simplified definition:

Non-obviousness—the innovation you've made that's different from what already exists is something that *would not be obvious to someone else in the field of your invention.* That is, if you isolate the part of your invention that is actually different from the prior art, would it be obvious to others in the field that they could have modified it in the way you did?

To understand what might be obvious, let's look at something that would definitely be considered obvious. If, for example, your idea is to make clothes hangers in smaller sizes for children's clothes, it may be that your invention is novel because no one had ever made a clothes hanger in exactly that size. However, it's probably the type of thing that people in the field of manufacturing clothes hangers know they can do. They know they can make a hanger in any size, including small sizes for children's clothes. They've probably chosen to make the conventional, standard size because it fits most clothes. The idea of changing the size of clothes hangers would therefore be obvious to people in the field of the invention, and it would not be a patentable idea.

Most of the time, changing the size, shape, color, or materials of an invention would be considered obvious by the Patent Office, unless there is an *unexpected*

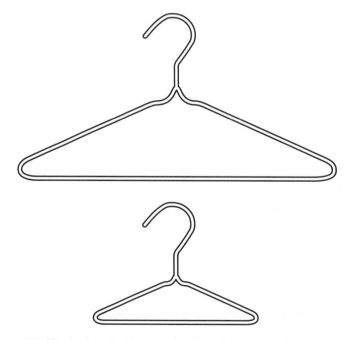

Figure 2.2 Merely changing the size of an invention is generally considered obvious

result. If, however, changing the material (or size, or shape) of an invention *does* provide an unexpected result, it could be patentable. For example, if you decided to make a telephone handset out of spongy material, and it provided the unexpected result that it actually improved sound quality on telephone calls, that might make your idea non-obvious and thus patentable.

Often, an invention is found to be obvious as a result of its existing in two or more parts in the prior art, if it would be obvious to people in the field of the invention that they could put the parts together and come up with the same thing that you did. An example might be something like a voice-activated video camcorder.

To evaluate whether it's obvious, we first need to look at the prior art. In doing so, we discover that voice-activated sound/cassette recorders already exist that start and stop *sound* recording when a voice is detected. And of course, video camcorders exist. So, the difference between our invention and the prior art is that we are proposing using this technology on a camcorder—which is a new idea because no one has recorded both sound and video via a voice-activated device before. Now, because the idea already exists for cassette recorders, this "new" idea of a voice-activated camcorder would not be a huge leap of logic. Just because the camcorder also simultaneously records video does not add anything unexpected. Thus, this invention would probably be considered obvious.

Sometimes it's said that "obvious" means people in the field *would know* they could do it, regardless of whether they have *actually chosen* to do it or have described it in some type of prior art document.

Another way of considering obviousness is to think about whether your idea is something that was "bound to happen" within its field. Most innovations that may be new but are seen as *inevitable*—given technological advances and where things are headed in its field—would be considered obvious. When the MP3 player first came out, it gave people the ability to carry around their whole music library stored on a tiny memory chip in a portable device. Many people thought, "Hey, what a great idea!" But to those within the industry, there was nothing groundbreaking about the idea itself. For a long time, they knew that a pile of CDs and cassettes was not the most convenient way for consumers to carry around and play their music. The technology to store music on a microchip existed for decades. It was not previously practical to do this on a large scale, however, because at the time it would have been way too expensive. I mean, in 1994, the amount of memory needed to store a single song would have cost about $200! But once the price of memory came down, having an affordable device that could store all of your songs on a memory chip was not just feasible—it was inevitable. And what might have looked like a life-changing invention to consumers was actually an obvious and inevitable evolutionary step within the field. That's how obviousness is judged—whether it would be obvious to a person in the field of the invention.

Now that we've seen some examples of what is obvious, we can begin to grasp what would be required for our invention to be considered non-obvious by the Patent Office. Generally, it means that there is something clever and unexpected about the invention, beyond what is shown by the prior art. Don't be intimidated by this! According to the records of issued US patents, there are currently more than 9 million examples of things that an examiner at the Patent Office considered to be non-obvious.

If you're confused about what is obvious verses what is non-obvious, then you're not alone. It is rather subjective. Frequently, this is where patentability gets confusing for the untrained eye. Even then, people can (and do) differ in opinion on the subject. For any given invention, the examiner at the Patent Office who reviews your patent application may see it differently than you or I. This is when an experienced patent attorney can give you guidance about whether your invention is likely to be considered obvious or non-obvious.

3. Can Your Invention's Operation Be Explained? (Enabling Disclosure)

The third requirement, *enabling disclosure*, is typically reserved for your patent attorney to worry about when he or she files your patent application. However, it's worth considering here because you want to make sure your idea is the type of thing that can be enabled. This means there must be enough information in your patent application for another person—one who is knowledgeable in the field of your invention—to be able to implement your idea just by reading the patent application and making or doing the thing. While you don't need to have built a working model of the invention in order to patent it, there shouldn't be any doubt from reading the application that it would be technologically feasible to do so.

In cases where it's unclear how the invention would/could operate and fulfill its purpose, it's doubtful whether even the most skilled patent attorney could enable it. For example:

> You want to patent a machine that reads your mind and knows what you're going to do next. It may be patentable subject matter. It may also be novel and non-obvious—but with current technology, no one could explain how such a device would work!

The important question to answer at an early stage, to help rule out unpatentable inventions, is: could *anyone* provide an enabling disclosure for this invention?

If there's no way to describe how it works, it's not patentable *because it's impossible to provide an enabling disclosure.*

Not sure how to fully describe your idea? Don't worry! Even when the inventor doesn't have all the nitty-gritty details of how an invention looks or works, a registered patent attorney, having both technical knowledge and expertise in writing patent applications, is accustomed to providing enabling disclosures for his or her client's ideas. If you have an idea for an app but don't know how to write code, that's okay—your patent attorney could likely explain how your app works with enough detail to provide an enabling disclosure. If you have an idea for a tractor that kills its own engine when a radio signal is received by a remote transmitter, and you don't know how to build such electronic circuitry, that's okay—because the technology exists to do it, and your patent attorney could provide an enabling disclosure.

But if your idea is to teleport people from place to place, like in *Star Trek*, it can't be patented because neither you nor your patent attorney could point to technology that makes it possible; therefore, you can't provide an enabling disclosure!

A rule of thumb you can follow is this: If there are people who could build your invention for you once you explain what the idea is, then it's probably possible for you and your patent attorney to provide an enabling disclosure. But if it's the type of thing that others would not know how to build, then no patent attorney could help you patent it! Thus, as a threshold consideration, you must ask, *could the operation be explained by present technology so that we can provide an enabling disclosure?*

Certain inventions require a technological leap to make them functional, or the design exposes a "gap" that no one could fill. Here are a few examples:

- *"I want to patent an antigravity device."*—This can't be patented because neither you nor anyone else knows how to make it work.
- *"I want to patent a smartphone app that senses your level of excitement while you are entering to-do items, and then it automatically schedules the items you are least excited about in the second quarter of your work day."*—The issue here is, how does the app sense your level of excitement? At least one example of how this can take place must be explained in your patent application, or there will be a gap in providing the needed enabling disclosure. However, keep in mind that this example need not be optimal!

Perhaps you theorize that when you measure the time it takes for someone to type in an entry, the slowest entries are the ones he or she is least excited about. It's probably a very weak theory, but if you assert this in the patent application, and explain that this is one example of an algorithm that the app might use to determine how excited the user is, it will probably suffice for providing an enabling disclosure! The example you provide *need not be the most workable*; it just needs to be workable. (There is, however, another requirement called *best mode*, which entails setting forth the best way that you know of for your invention to function, but I'll leave that to your patent attorney.)

- *"I want to patent a beer cooler chest that has a television built into the side panel."*—You might not know how a television works, but there are plenty of people who do. You would not need to describe the workings of a television in order to patent this combination, because it's well known. Accordingly, it would not require "undue experimentation by those possessing ordinary skill in the field of the invention" to produce that part of the invention.
- *"I want to patent a smartphone app that keeps track of how much time you spend using the individual apps, and then emails you a report of the percentage of time spent using each app. But I don't know how to write software or how this would actually work."*—In this example, not knowing how it would actually work is, again, fine. Within the field, it's well known how to write apps of all kinds. It is further known how to write apps that could work in conjunction with the smartphone's operating system to determine app usage, and could generate an email containing a report. So, as long as the patent application is written logically, there would be little doubt from the Patent Office that an enabling disclosure is provided.

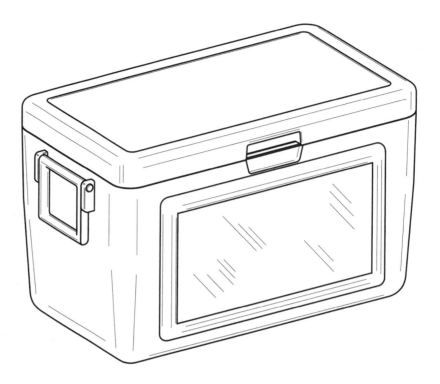

What this leaves out from patentability are inventions that are not currently technologically feasible or which contain parts that are not currently technologically feasible. It must be known how to *actually implement your invention*. That is, someone needs to know how to do it, whether that someone is you or others.

Summary

To be considered for a patent, your invention must fit one of the categories for patent protection: machine, manufacture, composition of matter, or process. To be patentable, an invention must be sufficiently different from previous inventions, known as prior art. In particular, it must be both novel and non-obvious. The invention must also be one that is capable of being explained in a patent application such that it is clearly feasible within present technology. For more information, visit www.patent-book.com.

Is Your Invention Really New?

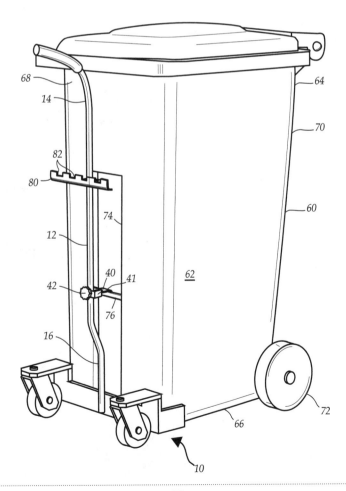

When you're considering filing a patent application, the first step is to figure out where you stand with regard to other inventions. To do this, you need to find the closest prior art to your invention. Examples of prior art can be found as actual products in the marketplace, and it can be written descriptions that have been published in various forms. It can also be in the form of patents—patents that exist going back to the very beginning of our patent system and those of other countries. Even *patent application publications*, which aren't patents but are patent applications that have been published by the USPTO or foreign patent offices, can be considered prior art. The point is, there's a variety of places where prior art may be lurking. And it pays to do your best to find out about the prior art before you invest in filing a patent application.

Search the Marketplace

Since existing products can be prior art, you should always start by searching for them. Take a look in all the places *you* would think to buy your product if it existed. You'll probably search the Internet for similar products, and you'll search in stores and in catalogs, places like that—places where you think it would be likely that you'd find the product.

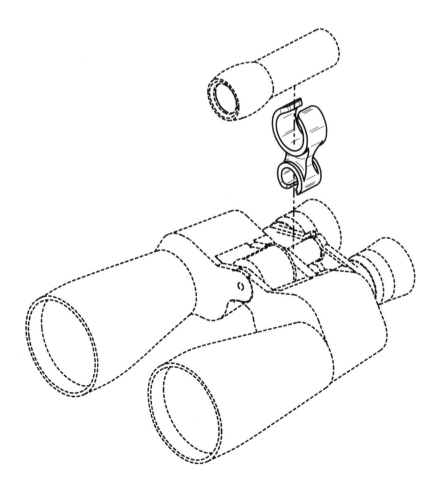

Patent Search

After you've searched in the marketplace and haven't found anything that's similar to your idea, next comes the patent search. When searching through existing patents, it's not so easy to find the closest prior art. Even expired patents—going back to the beginning of the patent system—are relevant to the *patentability* of your invention. Searching for the closest patents to your invention can be a really complex process, and it's often as much an art as it is a science.

What to Look for in Related Patents

When reviewing patents, many people get bogged down in trying to understand the strange language and structure of the patent itself. If you want to see whether an existing patent might be a barrier to your getting a new patent, it's simply a matter of:

- Does the patent describe and/or show your invention?
- Does it show the important features of your invention?

As you encounter patents that seem relevant, keep a list of them. You will want to show them to your patent attorney. The ones you find that you both believe are relevant to the patentability of your invention will need to be disclosed to the Patent Office when you apply for a patent. I'll talk more about this at the end of this chapter.

The General Idea vs. Your More Specific Solution

When searching for the first time, many inventors get discouraged if they see other examples from people who had the same general idea. With millions of patents in existence, it's quite expected to find that other people will have identified the same problem or need that inspired you, and they might even have thought of a solution that parallels yours. This is an opportunity to look more carefully at the specific features of your idea—at the solution you have devised and how it's different from the prior art you have found. When you look closely, you might find that your invention is a superior solution to the same need others tried to fulfill. And if the features of your invention are distinct, patenting it may still be possible and advisable. On the other hand, you might find some prior art with a more practical solution than yours and realize that perhaps it is time to move on to the next idea.

Online Searching

There are a few resources available on the Internet to help you find some of the patents that might be close in your field or close to your invention. This can be a good and inexpensive way to rule out pursuing an idea. If you can search for a short time, and you find that it exists, then it was well worth your time.

One very accessible and easy-to-use database is Google Patents. Go to http://patents.google.com, and you'll be able to search for patents the way you search for anything else through Google. It's not the most powerful or accurate searching tool, but it's certainly easy to use!

Be forewarned, however! Sometimes an online patent search (as through Google Patents) can give you a false sense of security. The search results may show you a few similar patents, which might give you just enough information to make you believe that you've found the closest prior art. But the truth is, there may be something closer that doesn't show up in these search results. And the patent that doesn't show up could be something that would get in the way of your getting a patent, or it could be something that would change your approach or the way in which you applied for a patent, if you'd only known about it.

The problem is that the various resources you can get to through the Internet are not really set up to find you the best prior art. Here's why: How do you typically search for things on the Internet? You use words, of course. You put in words that you think are probably going to describe a website that you're looking for. If you're looking for a burger joint, let's say, you type in something like "hamburger" and "restaurant." And the search results you get show a lot of websites that use those words. But when it comes to inventions—and in particular, patented inventions—there's a lot of different terminology that might have been used to describe inventions similar to yours. For example:

> Imagine that you have an idea for a can opener with a special type of handle. So you go to an online patent database, and you type in the words "can" and "opener." In some of these patent databases, you might come up with a whole bunch of can openers. And you might look through that list and see nothing there that's just like your invention. So you may think, "Hey, there's a big difference between my invention and the prior art, so I can go ahead and apply for a patent."

The problem is, what if someone had a can opener that had the same type of handle as yours, but that inventor didn't call it a "can opener"? Instead, she called it a "container opener." Or she called it a "device for opening metal containers," or even a "container piercing device." Her patent wouldn't appear in your search results when you searched "can opener" in an online patent database. This is one of the main reasons why, even when professionals try to do patent searches on the Internet, the results aren't really reliable enough to depend on when deciding whether to move forward with a patent application.

The bottom line: I recommend that you take the preliminary step of researching your idea online, because sometimes you will find exactly what you are looking for, which can save you time and money. However, if you don't find it in an online search, *you still can't be sure there isn't closer prior art*! While no search is perfect, the most effective way to search for prior art is to do what's called a *classification search* at the Patent Office.

Manual/Classification Searching

A classification search is a much more reliable way to find the best prior art, and it's the method most often used by professional patent searchers. In a classification search, you're not looking for inventions based on descriptive words, but based on the category of technology that such inventions fit into, according to the Patent Office's own classification system. The classification system includes more than 100,000 different class and subclass combinations. Each class/subclass has a definition, so that you know what it contains. For example, in class 7 (cutlery), subclass 400 is defined as "can openers." In a classification search for a can opener, a good starting point would be to look at the patents found in class/subclass 7/400. When you look at the patents in 7/400, if there is a similar patent to your idea—even if another inventor didn't actually call it a can opener, or used different terminology for the components—it would likely show up in the search. And depending on the particular aspects of your can opener, you would look further in the Manual of Patent Classification and find other promising places to search.

Do you want to know how the best searches are still performed? (Warning: here comes a bit of a long-winded explanation. If you are not interested, feel free to skip ahead to the next section.)

It used to be (just a few years ago) that you could go to the Patent Office Library in Arlington, Virginia, and search through paper copies of each patent. Even better, the patents were found in groups of other patents with the same class and subclass. You could pick up a pile of patents from that class/subclass and flip through them, quickly determining which ones were relevant. Then you could try another class and subclass that might lead you to the prior art you are looking for. In the end, you would check out several class/subclass areas before the search was complete. This was called a manual search, and it provided the best results. But alas, the Patent Office no longer has facilities for searching like this.

These days, professional patent searchers mimic the manual search by doing a "classification search." They use computer databases to retrieve patents within a class/subclass that seems appropriate and then review the documents online to see if they are relevant. The problem is that, generally, when you are looking through documents on the computer, it isn't as fast as looking at physical documents. How quickly could you flip through a 200-page book, looking at the pictures, to find the image you are searching for? Imagine doing that on a computer where you need to click "next page," and then wait for the page to load. It would take a lot longer, wouldn't it? Perhaps ten times longer than if you could flip through them by hand. The point is, the slower the document review process, the fewer documents a searcher can investigate, and the less ground that gets covered in a search. And good patent searches were always about covering as much ground as you could to see where that prior art might be hiding.

The solution: the USPTO still has search facilities (now in Alexandria, Virginia) that include computers connected *on their own intranet*. Because these computers have fast access to all document files contained on servers within the building, when you click "next page," the next page appears instantly each time—allowing searchers to look at more documents and cover more ground. Thanks to this faster access, many professional searchers still make the trip to the Patent Office to use their facilities.

By the way, your average patent attorney probably doesn't know or pay much attention to these details about the searching process. He or she orders a search from a search company and looks at the results when they come in. I happen to know this because my own brother, David Goldstein, is a professional patent searcher. And as he has explained to me, searching at the USPTO is still the best way to minimizing the chance of missing important prior art.

Have a Prior Art Search Done Before Applying for a Patent

As I mentioned earlier, when you get to the point where you've looked and looked and still haven't found your idea in the prior art, you absolutely want to have the right type of professional research done—including a classification search at the Patent Office—*before* you jump into the patent process.

You might find it interesting to know that some larger companies don't even bother with patent searches. They would rather get an application filed quickly and worry about the prior art later. But consider that these companies file many patent applications and aren't worried so much about the *win some, lose some* aspect of this approach. They are happy to spread their losses—from those applications that get shot down right away—across the numerous other patents they apply for and do actually receive. As an individual filing just a few—and perhaps just one—patent application in your lifetime, I doubt you want to take the "shoot first" approach. It would be better for you to do a search, and at least do your best to find out if you are wasting your time, before you apply for a patent.

Your Duty of Disclosure to the Patent Office

I'm getting a little ahead of myself by talking about this now, because we haven't yet discussed the *patent process*. But since we are talking about searching, there is something you should consider. As I mentioned earlier in this chapter, if you apply for a patent, you will have an obligation to reveal to the Patent Office any prior art that you know of that is relevant to whether your invention is patentable. This means that if you know of a patent for an invention just like yours, you must provide it to the Patent Office in an Information Disclosure Statement (IDS). If you submit an IDS when your application is filed, or soon after, it's free. If you submit it later, you must pay additional fees (beyond your application fee) to the Patent Office.

When to Stop Searching

It is certainly important to find and examine the prior art before you apply for a patent. This part of the process is a *patentability review*. It can help you decide whether applying for a patent is a good idea, and will help shape how your patent application is written.

But once your application is filed, you might want to stop looking for a bit. Some people can't stop themselves, however. They *oversearch* their invention when there is little or no benefit to doing so. I mean, even after they have had a patent application prepared and filed, they search, and they search, and they search—as if it's a full-time job—until they find the perfect grounds for rejecting their patent application. Then they send the patents they found to their attorney and ask, "Hey, what about this?" Now, their patent attorney is obligated to provide all of this information to the Patent Office. Remember that *any relevant prior art that you find* must be brought to the attention of the Patent Office examiner for your case!

Here's the danger: the information you submit might not be anything an examiner at the Patent Office would have ever thought relevant to your case. Frequently they are patents that bear some similarities but are not totally related.

However, when you hand them to the examiner in an IDS, of course the examiner will second-guess whether they might be important. And remember, too, that whether an invention is considered obvious in light of the prior art is open to a good deal of interpretation by the examiner. I mean, there's a lot of latitude in how the examiner might *connect the dots*, or in which dots the examiner would ever even think to connect. But if you hand someone a bunch of dots, chances are that person will find a way to connect them!

As a result, just sending these patents in will cause the examiner to give your application a second look. It's only natural that the examiner will wonder why you considered this prior art relevant to your patent application. All this conjecture may very well shape his or her opinion in favor of rejecting your application!

Why do people do this? I think they are trying to justify that they made a good decision to file a patent application. It's kind of like—before you buy a car, you might do research to find articles that rate and compare the cars you may be interested in buying. But once you go through with it and buy the car, the decision has been made. *It's over.* No need to do any more research! If you still have an itch to look up the car you bought and see how it stacks up against other cars, it's because you're now second-guessing your decision. The same is true about oversearching. When people relentlessly search for prior art after they are well committed to the process—after the work has been done—it's just about their own second-guessing. But in this case, that second-guessing could be harmful to the outcome of their patent application.

I want to make it crystal clear that you **must** submit any relevant prior art that you know of to the Patent Office. But once you're committed, once the application is written and filed, be cognizant of the potential harm of digging too deep when it no longer serves you.

Other Types of Research and Due Diligence

There are other types of patent research that can help you make decisions as you contemplate launching a product. Usually they become more appropriate further along in the product development cycle, as additional capital is invested in your business and as you move closer toward your product's final design and launch.

Infringement Evaluation

If you believe you might be infringing a particular patent or "family of patents", a patent attorney can perform an "infringement evaluation". The attorney will review the existing patent and compare it to your product or proposed product, to determine whether your product is likely to infringe. It can be a time consuming process for the attorney, and thus can be expensive—even when considering just a single patent.

Freedom to Operate Opinion

When a company is about to launch a product, the company might request a "freedom to operate" opinion. As the name implies, its purpose is to see whether there are any patents that might be infringed by the product. This is a much broader inquiry than a patentability review, since it's not just a question of whether your key combination of features is unique, but also a question of whether any of the subparts might infringe another patent. In terms of labor intensiveness, freedom to operate opinions are a few steps beyond infringement evaluations.

The main difference is that potentially infringed patents have not yet been identified, so the attorney, and the researchers employed by the attorney, will first seek to find patents where a potential infringement issue might exist. The attorney will then review all of these patents in detail to see if infringement is likely.

Design Around

When you know of a patent that covers a product you would like to manufacture and sell—in other words, when you discover that the main idea of your product really isn't new—you might engage in a "design around" inquiry to see how you might get around the patent. Usually, a design around is less formal and more interactive with the attorney. The attorney will analyze the existing patent and let you know what features are required to create infringement of that patent. Then you can begin to consider whether those features are necessary for you to produce a viable product.

As a general rule, if you can leave out features that are required by the "patent claims" of another patent (more on this later), you might be able to safely manufacture your product. The caveat is always: Just because your product is *technically* not infringing doesn't mean your infuriated competitor might not try to sue you anyway!

Summary

Prior art searching is typically performed before making the decision to apply for a patent. Search online first to determine if your invention clearly is not new. A more thorough professional search should then be performed before actually committing to the patent process. Other types of searches and due diligence might become appropriate later on, as you ready a product for launch. More information about patent searching can be found on the companion site for this book at: www.patent-book/patent-search.

How Can Filing a Patent Application Protect Your Invention?

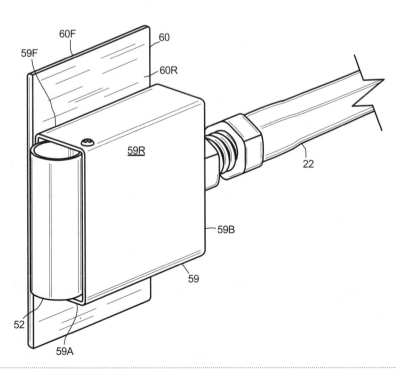

Often people ask: "If we file this patent application, will I be protected?"
My response: "It depends on what you mean by 'protected.' Let's talk about it and I'll explain."

First Thing to Know: Filing a Patent Application Protects Your Priority Toward Getting a Patent

When your patent application is filed, a *priority date* is established. Your priority date is used not only for determining what prior art the Patent Office may use in judging your application but also for determining who is entitled to the patent. Under our current *"first-to-file" system*, if after you have filed, someone else files a patent application for the same invention, you will have priority over him or her for getting the patent because of your earlier priority (filing) date.

Filing a patent application, then, protects you in the sense that no one filing after you will be able to "beat you out" to get the patent. Thus, once you have filed, if you talk to people about your invention, you don't need to fear that they might like the idea so much that they try to get the patent ahead of you. Because your patent application is already filed, you will have priority ahead of them. As a practical matter, this is the protection that means the most to inventors at an early stage. They want to establish their rights sufficiently so that they can talk with people in the field to get feedback, gather information, and possibly find a venture partner. At this stage, they are not as worried about being copied (because that will take time). They are more concerned that someone else will try to get the patent. Having a patent application already filed helps inventors feel protected against this happening.

IMPORTANT CONCEPT

Not all patent applications are created equal.

The priority that is established for you—as the first inventor of the product—is only as good as the description used in your patent application. And a patent application is only as good as the quality of the writing within it. The quality of the patent application, be it utility, provisional, or design, is critical to getting a strong patent, and even to getting any patent at all. A poorly written application might have an earlier date than a competitor's application, but is far less likely to result in a strong patent, one that you could use to stop people from stealing your invention.

The one thing that truly has a chance of getting you strong patent protection for your idea is an application that's been written correctly, with: (1) full understanding of what your invention is, (2) full understanding of what's distinct about your invention, and (3) full understanding of your goals for this invention. This is an important prerequisite toward ensuring that the patent you end up with is a match for your needs.

Second Thing to Know: You Need an Actual Patent to Protect Against Copying

To stop others from making, using, or selling your invention, you need more than just a patent application—you need an actual patent. Every patent starts with an application. Once the application has been reviewed and approved by the US Patent Office, it issues as a US Patent. Only at this point can the patent be enforced to stop others from making, using, or selling the invention that it covers.

Third Thing to Know: The Type of Copying You Can Stop Depends on the Patent

You'll often hear people say, "It's patented, so no one can copy it!" This is a bit of an oversimplification. Whether someone can copy it depends on the *type* of patent and the *scope* of the patent.

In a moment, we'll talk about two main types of patents: *utility* patents and *design* patents. They each protect different types of things—different aspects of a product. Also, every patent varies in scope. That is, some patents are *broader*—effectively covering an entire concept—while others are *narrower* and more about the details. When we talk about patent claims in Chapter 6, and about infringement in Chapter 15, you will better understand what it takes to obtain a broad patent and how to determine the scope of protection provided by your patent.

Utility Patent

If the invention is a combination of structural elements—a combination of pieces that are put together in a unique way for a functional purpose—then a utility patent is probably the way to go about protecting it. Utility patents can also be used to protect a method, which is a combination of steps that are arranged for a functional purpose to carry out a desired functional result. A utility patent can be the most appropriate form of protection to prevent other people from making, using, or selling the same technology, the same machine, or the same idea as what you have in mind.

A utility patent application should illustrate and describe a workable example of your invention to the satisfaction of the rules of the United States Patent and Trademark Office. It also defines the boundaries for the invention—what others cannot make, use, or sell without your permission.

Design Patent

A *design patent* protects the ornamental appearance of a functional object. When the aspect that makes an invention special is its shape—for ornamental purposes and not in service of its functionality—then a design patent can provide the appropriate type of protection. If the distinguishing point of your invention is something other than its shape—something specifically related to its functionality—then a design patent probably isn't the best *primary* form of protection, but it could be a good *secondary* form of protection. If what's different about your idea is functional, then ultimately, it's a utility patent that you'd be looking for.

For many inventions, however, it's the design itself—the ornamental appearance—that represents the most significant innovation. In that case, obtaining a design patent might be even more important, and more feasible, than obtaining a utility patent. Also, generally speaking, the cost of obtaining a design patent is significantly less than the cost of a utility patent.

Design Patent Drawings

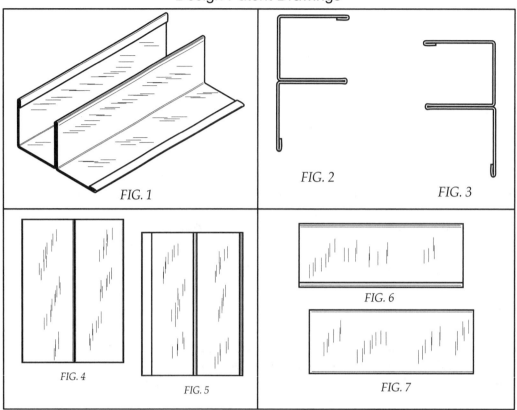

FIG. 1

FIG. 2

FIG. 3

FIG. 4

FIG. 5

FIG. 6

FIG. 7

Utility Patent Drawing

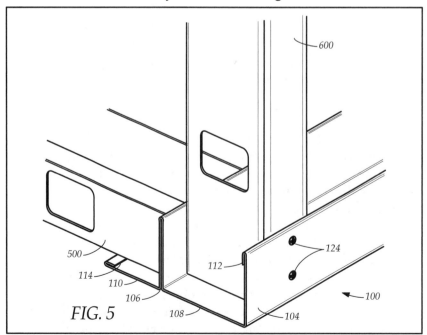

FIG. 5

Figure 4.1 Design and utility patent drawings for a product

Seeking Both Design and Utility Patents for a Product

Many products have both design and utility patents. The reason is simple: there are functional portions that are protected by the utility patent, while the shape and appearance are protected by the design patent. For any modern automobile, for example, there are hundreds and even thousands of patents involved. Innovations in the transmission operation, in emissions systems, or in driver instrumentation operation may be the subject of utility patents. Design patents cover everything from the innovative shape of the bumpers and headlights to the shape of the dashboard and center console.

For your product, it may be that the structure and function are suitable for a utility patent. Once you have a finished product, ready to ship, it is quite likely you will also want to file for a design patent to protect the curves and nuances of the product you are selling. In this way, the design patent will protect against "unimaginative copying." This is when someone picks up your product and, rather than using it as inspiration to design their own product following your concept, they instead just ask a manufacturer to "make me 10,000 of these." In that case, the design patent can be an effective weapon to prevent such copying.

I often file both design and utility patent applications for the same client. The strategy is: the utility patent would be more valuable but might be more difficult to obtain, while the design patent might be less valuable but easier to obtain. Hopefully, we get both patents! But if we don't end up getting the utility patent, most likely we will still get the design patent.

When you file both a design patent application and a utility patent application, each application will follow its own process. Each application is prepared separately, emphasizing different aspects of the product, as illustrated in FIG 4.1, and is filed separately. Each will require its own filing fee and will likely require separate attorney's fees. Once filed, each will be processed independently by the Patent Office. They will be assigned to different examiners in different parts of the Patent Office, and each will wait its turn to be reviewed. When a determination is made about patentability, the determination will be made according to different criteria, and often in light of different prior art. Lately, design applications have a much shorter wait than utility applications to be reviewed by an examiner. Chances are, your design application will be reviewed and issued well before an examiner even takes a look at your utility application! When that happens, in the meantime, you can mark your product "US Patent Des. 123,456. Other patent(s) pending."

What Is a Provisional Patent Application?

A provisional patent application is a stepping stone toward a utility patent. It establishes a priority date for you that you will keep as long as you file a utility patent application within the next year. Sometimes, it is used as a lower cost alternative, to buy some additional time before filing a full utility patent application. Other times, it is used when the invention is likely to develop significantly in the coming months but you want to establish your rights as early as possible for what you've already figured out.

Once again, the point of the provisional application is to establish priority for you. When the provisional application is filed, you establish a documented filing date and can label your idea as "Patent Pending." But the provisional application

IMPORTANT CONCEPT

What about an NDA instead of a patent?

A non-disclosure agreement (NDA) is a contract in which someone promises not to disclose your idea to others. You might be thinking, "Why don't I just have everyone I'm working with sign an NDA? Then I won't need a patent, right?" Not so fast. Because an NDA is a contract with *only* the person (or company) who signs it, it does nothing to prevent third parties from copying your idea. If you have a patent, however, even people you have never dealt with could not legally make, use, or sell your invention without your permission.

will not be reviewed by the Patent Office, and a patent will not be granted from just filing the provisional. If you file a utility patent application within one year of filing the provisional application for the same idea, you'll get the benefit of priority for your claim of ownership of the idea from that first provisional filing date.

In Chapter 8, we'll discuss the advantages of filing provisional applications in more detail.

Plant Patents/Patenting Genetic Material

Certain biological inventions or innovations also may be patentable, such as a plant or a piece of genetic material that has been isolated or developed in a lab or other controlled environment. While genes and most plants may indicate filing a utility patent application (as a composition of matter), the patent laws provide for a separate type of patent, known as a "plant patent," to cover certain asexually reproduced plants, including some types of fungi and algae.

The general rules of patentability—patentable subject matter, novelty/non-obviousness, and enabling disclosure—also apply to these types of inventions. But there are additional requirements that must be met. For example, a plant that was not intentionally developed still may be patentable if it was discovered in a cultivated field, greenhouse, or other controlled area. If it was discovered in the wild, it is not patentable. Additionally, the plant's distinguishing characteristic(s) cannot be a result of something as simple as using a different fertilizer or varying its exposure to light or water.

For genetic material to be patentable, it must be rendered in a purified state, either by separating it from other naturally occurring compounds or by synthesizing it from its component chemicals. In addition, the material must have a specific utility—it has to be useful for *something*, and that use must be spelled out in detail in the patent application.

Summary

The way to begin establishing your rights is to file a patent application. A utility patent application may be filed to protect functional improvements, and a design patent application may be filed to protect the ornamental appearance of a useful item. For many products, it pays to apply for both. A provisional patent application can be used to temporarily establish priority for a functional invention. For more information on filing patent applications, visit www.patent-book.com/file-a-patent.

How Does the Patent System Work?

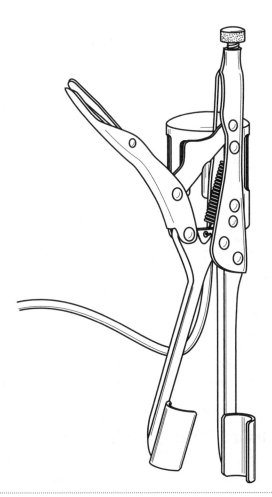

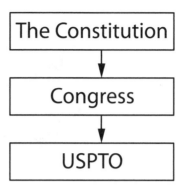

"The Congress shall have Power . . . To promote the Progress of Science and useful Arts, by securing for limited Times to Authors and Inventors the exclusive Right to their respective Writings and Discoveries."

Our patent system has its roots in the US Constitution. The reason we even have a patent system is that the framers of the Constitution thought it wise to provide legal protection for innovations. In particular, Article I, Section 8 of the US Constitution mandates that Congress make laws to promote progress by protecting the works of "Authors and Inventors" for a limited time.

This is useful to know because it shows that, while protection is fundamental to our patent system, it is also fundamental that the protection will be provided only for a limited time, which is why patents expire and cannot be renewed. It also shows that the intention was to promote progress, which requires that the works be openly shared, and that they will eventually enter the *public domain*.

As this Constitutional framework relates to patents in particular, the "bargain" that is at the core of the patent system is that you, the inventor, fully describe your invention, so that others can understand what you've created and, when the patent expires, can freely practice it and build upon it (to promote progress). In return, the government will grant you protection for a specified period of time.

Our Patent Laws Come from Congress

In fulfilling its duty under the Constitution, with the Patent Act of 1790, Congress established our first laws governing how patents would be considered, granted, and enforced. In the 200+ years since then, Congress created the United States Patent and Trademark Office (USPTO) to carry out examining applications and granting patents and has periodically enacted additional laws that have brought the US patent laws to where they are today. Most recently, the America Invents Act (AIA) brought sweeping changes to the patent laws, which included instituting a "first-to-file" system in the United States, for the first time in its history.

The United States Patent and Trademark Office

The USPTO is the organization responsible for receiving US patent applications, reviewing them, and granting US patents. The Patent Office also makes rules and

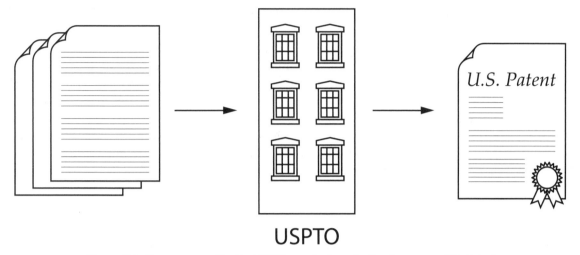

Figure 5.1 Once approved by the USPTO, a patent application becomes a U.S. Patent

procedures regarding the examination of patent applications. These rules must be consistent with the laws enacted by Congress.

Federal Courts

Federal courts have the power to hear patent infringement lawsuits. If someone is infringing your patent, you can file a lawsuit in federal court against the infringer. Federal courts have exclusive jurisdiction over patent infringement lawsuits.

Many people falsely believe that if someone is infringing, you go to the Patent Office to complain about it. It's important to understand the distinction between the roles of the Patent Office and the courts: the Patent Office grants the patents; the courts enforce them. But note that final decisions of the Patent Office can be appealed in the federal courts—in some cases, all the way to the Supreme Court. And also note that the Patent Office has procedures for challenging the validity of existing patents, some of which can resemble a lawsuit in a court of law. Nevertheless, the appropriate place to address infringement of your patent is not at the Patent Office.

The Process for Obtaining a Patent at the United States Patent Office

Once you and your attorney have determined that what you have is novel, is non-obvious, and is worthy of a patent, the next step is to file a patent application in the USPTO. But first, the application must be *prepared*.

Preparing (or "Drafting") the Application

Preparing a patent application means getting together information about what the invention is, what's different about the invention, and why it's worthy of a patent. A typical application can be twenty to forty pages long. The point of all this documentation is to firmly establish what your invention is and to begin the process of convincing the Patent Office to grant you the patent.

We established at the beginning of the book that you are not under the misconception that applying for a patent is something you can do on your own. If you still have some "do-it-yourself thoughts" about this, I want to emphasize the following: There might be parts of the application process that you can do yourself—and that you should do yourself, such as an early prior art search. But writing a patent application is the most critical part, the most technical part, and the part that requires the most skill of anything in the entire process.

Filing the Application

Assuming that your application has been prepared and is everything that it needs to be, it's then filed in the US Patent Office. On the day that it's filed, your idea becomes *patent pending*. You've probably seen products marked as "Patent Pending." Basically, this means that the application has been filed, and the inventor is waiting for the Patent Office to review the application and grant the patent. When you're in patent pending status, it becomes a bit of a waiting game as far as getting the patent itself. You're waiting for the application to be *examined*.

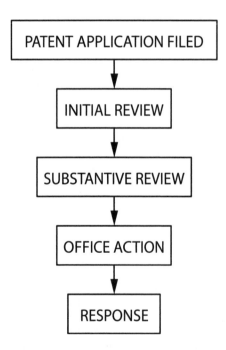

Initial Examination

After the application is filed, generally within a few weeks, it will be given an initial review to make sure it is complete. At this point, the folks at the Patent Office are not considering the merits of your invention; they generally are just looking to make sure all pages and drawings are present, whether supporting paperwork, such as the oath/declaration of the applicant, was included, and whether the proper fees have been paid. If something is missing, they will issue a *notice of missing parts*.

When there are only minor defects, the notice of missing parts will indicate that a filing date was granted. In cases where there are more serious defects, the notice will indicate that a filing date was not granted. In either case, it will provide instructions for responding and a deadline for response.

If the notice was issued because of missing information, you'll have the opportunity to get the rest of the information to the Patent Office. If it's rejected for more serious reasons, you may be able to argue the point or amend the application, depending on the reason for the notice.

Also during this initial review, a determination will be made whether anything in the application threatens national security. This is done just to make sure patent applications that contain critical secrets aren't filed in foreign countries. In theory, if the application did contain such information, a secrecy order would be made. This is practically never the case, so don't worry! What they do instead, at least 99.9 percent of the time, is determine that there are in fact no critical secrets in your patent application and you grant a *foreign filing license*, which then gives you the go-ahead to file patent applications in foreign countries, if you wish.

When your application is determined to be complete, a filing receipt containing a foreign filing license is generated and sent to you, and the application is forwarded on down the line. It then will be given a *classification* (determined by the type of invention it is) so that it can be assigned to the correct examining group. The Patent Office has numerous *examining groups*, which are arranged by technology area, and each group employs numerous *patent examiners*. For example, one group might examine certain types of mechanical inventions, while another group examines chemical compounds, and yet another examines software inventions. Once your application is sent to the appropriate group, it will be put in line to be examined.

Some groups and some examiners are busier than others and have a different-sized backlog of applications waiting to be examined. Depending on the staffing, work output, and new patent application filings each year, the average wait time increases and decreases. In general, however, you can expect to wait approximately one to three years before your application is "substantively examined."

Substantive Examination

As noted previously, after the initial examination, your application will be assigned to a Patent Office examiner whose job is to review your application, do his or her own prior art research, do his or her own comparisons, and make his or her own judgment as to whether your invention fits the criteria for patentability.

As we previously discussed, provisional patent applications are not examined, nor do they result in an issued patent.

This *substantive examination* is more in depth than the initial examination. The patent examiner digs into the details of the claims section of your application, comparing it to all applicable prior art that you disclosed with your application, as well as any that the examiner has found.

Technically, what the examiner is doing is looking to find prior art that fits the description of the claims in your patent application (see the section on claims in Chapter 6). The examiner will compare the prior art to your claims to determine whether your claims describe an invention that is patentably distinct from the prior art, using the criteria of "new" and "non-obvious" as we discussed in Chapter 2. The examiner also gets into the nitty gritty of your disclosure, to determine whether it provides an enabling disclosure, i.e., someone with knowledge in the field really could duplicate your idea from the information you have provided. Additionally, the examiner will review the application to see if it violates any of the myriad Patent Office rules about how an application is supposed to be written and how the drawings are supposed to be presented.

Sometimes the examiner's review isn't that favorable, and he or she might then issue a *non-final Office Action* rejecting the application. The reasons for the rejection are always provided in writing so that you and your attorney can review the reasons and determine whether the rejections can be overcome.

Receiving an initial rejection is fairly common. The majority of patents were first rejected before they were approved. In practice, the majority of rejections can be overcome—either by disputing the examiner's reasoning on the issue or by amending the application. Of course, rejection is not fun, and it will add to the expense, but it's part of the process.

Often an Office Action contains both *prior art rejections* and *objections* to minor "matters of form." Sometimes—more frequently with software-related inventions—the examiner may make *subject matter rejections*, and contend that *patentable subject matter* is lacking. For more information about Office Actions, what they may contain, and ways to overcome them, see Chapter 9.

When making prior art rejections, the examiner will refer to prior art references, which are typically patents and publications. The examiner will compare these references to your claims and indicate how, in the examiner's opinion, these patents and publications render your invention unpatentable. In doing so, the examiner will either state that the references show the *exact* invention defined by your claims (that it lacks *novelty*) or that they are not exact but make your claimed invention *obvious*. Since most patent applications contain several claims, most Office Actions will contain several rejections, often pointing to different prior art references and combinations of references and stating that they render the various claims unpatentable.

Objections concerning matters of form, how the application is written, and so on, may spell out various ways in which the terminology used in the application is inconsistent. For example, if a part was called a "window frame" in the claims but was called "window framing" in the rest of the application, this will lead to an objection from the examiner. Also, when it is unclear how an invention will function, or if the description contains significant gaps, this may lead to a more significant rejection.

Subject matter rejections contend that the invention is not the right type of invention that can be patented. As discussed in Chapter 2, all inventions must fit at least one of the four categories of patentable subject matter. A rejection stating that patentable subject matter is lacking is most commonly given for software-related inventions (see Appendix A for a further discussion about software patent applications).

Response

Most *responses* are typically a combination of *arguments* and *amendments*, in an attempt to convince the examiner to remove the rejection. Generally, you are given three months to respond, with an additional three-month extension possible by paying extension fees. Drafting and filing a response can be time consuming, so expect that additional professional fees will be required at this point.

Some Applications Receive a "Restriction" Before Substantive Examination

Only one invention is allowed per patent application. When your application is written, however, most skilled attorneys try to include different "approaches" at defining your invention. Sometimes this means including some claims that refer to the device itself, some that refer to how the product is manufactured, and possibly even a method of use.

Before substantively examining a patent application, as we discuss earlier, the examiner will sometimes decide that these different approaches within the same application can be considered separate inventions and will then issue a *restriction*. The restriction requires that the inventor pick one of these inventions to be examined now. In response, you select one by filing an *election*, in which you choose the invention you wish to have examined. When you file the election, you may sometimes also argue that the restriction by the examiner was improper. In any case, as illustrated in Figure 5.2, the invention you don't pick now can be pursued later in a *divisional* application and can result in an additional patent (more on this in Chapter 9).

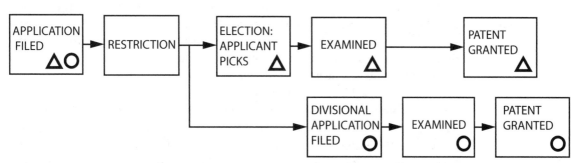

Figure 5.2 Following a restriction by the examiner, the non-elected invention can be filed as a separate "divisional" patent application

Final Office Action

If your responses to any of the examiner's reasons for rejection do not overcome his or her objections to your application, the examiner will issue a *Final Office*

Action. This rejection letter states why the responses you provided weren't sufficient to overcome the examiner's earlier objections. As "final" as it sounds, a Final Office Action does not have to spell the end of your patent journey. You can still submit another response, provided that it *does* overcome the examiner's objections, or you can pay a fee and file a *request for continued examination* (RCE). After receiving a final rejection, you also can appeal the decision to the Patent Trial and Appeal Board (PTAB). An appeal provides the first opportunity to have a board of experienced examiners consider whether your examiner was correct in rejecting your application. And if you are not satisfied with the result of this appeal, you can then take further action in federal court. Appeals are not that common, especially for patent applications filed by individual inventors. They take a long time to be heard and decided, and the expenses can rack up quickly.

Abandonment

Sometimes it doesn't pay to respond. Perhaps the examiner's rejection seems insurmountable, or your circumstances have changed and the patent application is no longer relevant or no longer a priority, or, in the balance, pursuing it further doesn't seem worthwhile. Other times, you may have filed another patent application that has a more promising outlook for approval, and it makes sense to *abandon* this one.

If you fail to respond within the permitted time period, your application will be considered *abandoned*. Generally, this is the end of the road. In certain circumstances, however, you can petition the Patent Office to have the application revived. Doing so typically requires paying significant petition fees and can only be done for a relatively short period of time after abandonment. In addition, you can only revive an application when it was abandoned by mistake or if abandoning it was unintentional or unavoidable. Thus, you can't revive a patent application if you decided to abandon it but later changed your mind and decided to pursue it.

Notice of Allowance

Assuming that the review by the patent examiner is favorable, you'll be issued a *Notice of Allowance*. When the Patent Office issues this notice, it's essentially saying that your application has been approved for a patent. After you then pay the appropriate *issuance fees*, the United States Patent will be issued, and a formal patent certificate will be printed and mailed to you.

In practice, a Notice of Allowance won't likely be the first result of substantive examination. Typically, as we discuss previously, you will first receive an Office Action containing a rejection. After you respond, and if your response overcomes the reasons for rejection, the examiner typically will remove the rejection and issue a Notice of Allowance. Then, once you pay the issuance fees, the patent will be granted.

Allowance Following an Examiner's Amendment

Sometimes, when an examiner sees that the application is very close to being ready to allow, the examiner will call the patent attorney representing the inventor and suggest a few changes that would put the application in condition for allowance. If the patent attorney agrees, the examiner will enter the changes by *examiner's amendment*, and then issue the Notice of Allowance.

Payment of Issuance Fee

You have three months from the date of the Notice of Allowance to pay the appropriate issuance fee. It is important to understand, however, that this time frame *cannot be extended*. If you don't pay by the deadline, the patent application will be abandoned.

Patent Issuance

Once the patent is issued, you have the right to exclude other people from making, using, or selling your invention within the United States. This is a right that you can keep to yourself, sell to someone else, or license to a company that wants to make your product or use your IP in its own products. With a utility patent, your rights will last until the patent expires twenty years from the filing date, as long as *maintenance fees* are paid (discussed below). For design patents filed after

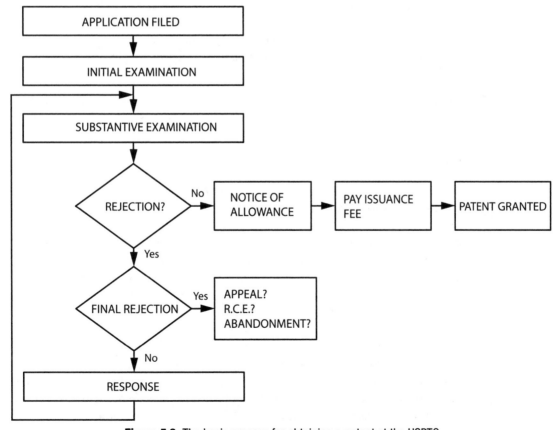

Figure 5.3 The basic process for obtaining a patent at the USPTO

May 2015, the patent expires fifteen years from the issue date, and no maintenance fees are required.

Maintenance Fees

US utility patents require that the inventor pay maintenance fees to keep the patent in force. These fees must be paid by the four-, eight-, and twelve-year anniversaries of the patent issue date. If they are not paid, the patent will expire early.

Speeding Up the Process

There are mechanisms to give your application special status at the Patent Office, and thereby reduce the time you wait for examination:

- **Petition to Make Special**—You can file a petition to have your application granted special status. When an application has special status, it will be examined ahead of turn. In some cases, this reduces the wait time to just a few months. There are various grounds for attaining special status. One is the "advanced age of the applicant." What this means is that if you are over age sixty-five, you can get your application granted special status. Other grounds include ill health of the applicant, when infringement is already present, if the invention advances energy conservation, and so on.
- **Track One**—The Patent Office has a program whereby you pay a surcharge to speed up the process. Under this program, their goal is to bring the application the whole way to either a granted patent or a Final Office Action within a year. Typically, this means that the application is examined in just a few months.

Depending on your circumstances, either of these programs could greatly reduce the wait time for examination of your application and issuance of your patent, should you—and your idea—qualify.

Is It Worth Speeding It Up?

Naturally, it seems like faster would be better. When most people hear that it will take years to get a patent, they wonder what they can do to speed things up. Sometimes inventors are hesitant to do anything toward commercializing the idea until they know for sure that they will get the patent. Other times, inventors want to license their idea and the other party wants an issued patent before making a licensing deal. But consider that speeding it up will (in most cases) also mean receiving that first Office Action much faster, which will probably mean the expense of responding to that first rejection will be incurred much sooner!

I find that a lot of inventors prefer a little time/space after filing their application to get things rolling, and see how it is going with the product, before they are confronted with the decision to either spend more or abandon their application. Also, allowing some time can provide additional perspective—and sometimes even evidence of success—that may be helpful in responding to a rejection. Further, while it seems useful to actually have the patent in hand in case someone is infringing it, most of the time the infringers don't come along until much later.

As a practical matter, it's relatively rare for the infringers to come along while the patent is still pending. As a result, other than cases where the inventors refused to do anything at all until they had the patent in hand, or when a patent was needed for licensing, I have rarely seen cases where getting the patent quickly actually mattered. Consider carefully, then, whether in your circumstances you actually need the patent to issue faster, and whether it's really worth trying to speed up examination of your patent application.

Summary

The process of obtaining a patent starts by preparing a patent application and submitting it to the Patent Office. An examiner at the Patent Office will review your application and either approve or reject it. If it's rejected, you'll have an opportunity to respond and seek to win approval of the patent application. Once approved, your patent application will issue as a United States Patent. For further information about the patent process, visit www.patent-book.com/patent-process.

What Goes into a Patent Application?

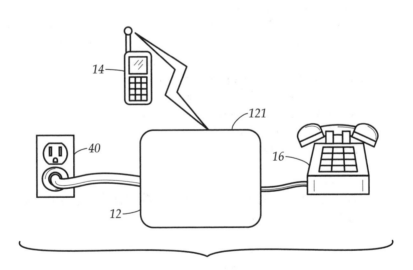

I think we are clear by now that it's not the intention of this book to teach you how to write a patent application. But I do want you to understand the considerations and strategy that dictate how your patent application is written and prepared.

The first thing to know is that, once a patent application is approved, the application essentially becomes the patent. Consequently, however the patent application was written will forever be the official text of your US Patent. And whatever drawings are provided will become the official drawings in the printed US Patent.

If you have ever wondered, then, why patent applications are time consuming and expensive to prepare, consider that it is at the drafting stage of your patent application that what will eventually become your patent is being prepared!

First we'll discuss utility patent applications. If you are curious about what goes into a design patent application, look to the end of this chapter.

In talking about what goes into a utility patent application, let's start with the biggest question I hear, and perhaps the most important concept you should understand about your utility patent application:

Should my patent application have a lot of detail, and will that limit my invention to those details?

I hear clients say all the time, "I want to get a patent on the concept, not on the details of how I do it," or "I want this to be patented, no matter what material someone uses to make it," or "Don't mention the fact that it is flexible, I don't want to be limited to that."

All of these clients are expressing the same thing: they don't want a patent that would only prevent someone from copying *their exact product*, or the very specific combination of features it contains. What they want is a patent for their basic concept, so they can prevent people from copying their concept. This is exactly what we (I am speaking for all patent attorneys here) set out to achieve for our clients. But the fear that providing too much detail will limit the scope of protection provided by the patent is misplaced. As long as it is handled correctly, additional detail in the application will *not* limit the scope of the patent. I'll explain below.

The "Specification" and the "Claims"—(How Much Detail to Include and Where to Put It)

The majority of your utility patent application is known as the *specification*. It is also sometimes called the "spec" or "the disclosure." The point of the specification is to fully explain the invention. There is no need to censor yourself or delete details because the whole point of the specification is to provide as much as you know—no holds barred—to explain the invention to others. The details are of course presented in a logical, well-organized, hierarchical fashion.

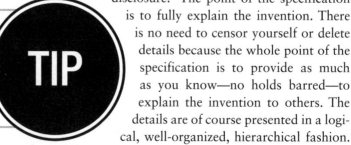

Just mentioning a feature in your patent application won't necessarily limit you to that feature!

TIP

A well-written specification connects the dots among all of the key components of the invention, and teaches how the invention is structured and how it is operated and used.

The *claims* accompany the specification in your patent application. In contrast to what we just said about the specification, in the claims, less detail is actually better! The claims section of the patent is about precisely defining the invention as simply as possible. In the claims, you establish the definition of the invention that will later be used for determining infringement. Every word is critical! A claim that contains less detail will be *broader*, which means that it covers more and will make it more difficult for others to copy your concept. To a great extent, then, how strong your patent is depends on how the claims are written. I will explain how claims work in detail later in this chapter.

The Odd Language and Structure You'll Find in a Patent Application

Throughout the application, the way it's written—the language and the structure—may seem quite strange, especially if you haven't read many patents or patent applications! But the Patent Office has very strict rules about how an invention must be described.

As an example, one such rule requires that we introduce each new element (each part of the invention) in relation to the other elements. As we introduce each element, we establish *antecedent basis* for the next time we mention it. Following this rule, every new introduced element starts with "a" or "an," and then you can use "the" or "said" the second time you refer to it. For example, a passage you might find in a patent application may read:

> "A toothbrush, for use by a user having a mouth with a plurality of teeth, and at least one hand having a thumb and forefinger. The toothbrush has a handle, and a brush head attached to the handle, the brush head having bristles. In use, the user holds the handle between the thumb and forefinger to direct the brush head toward the mouth and engage the bristles with the plurality of teeth."

In real life, we don't talk like this! But in a patent application, this kind of wording is necessary if we want to talk about how the toothbrush is held by the user between his thumb and forefinger and used in his mouth to clean his teeth. We need to *establish context* in the patent application for the device and its elements, for where the invention is used ("by a user having a mouth . . ."), and then we can connect the two. If we fail to do this, and we begin talking about an element out of the blue—well, among other things, we might get a rejection for "lack of antecedent basis." Just one of the many oddities peculiar to patent application writing!

How a Utility Patent Application Differs from a Provisional Patent Application

The primary difference between a well-drafted provisional application and a utility application is that the utility application includes *claims*. The claims section carves out the territory that the utility patent will cover and sets the boundaries for what may be considered *patent infringement*.

Parts of a Utility Patent Application

The following components make up a utility patent application:

- Background
- Summary
- Brief description of drawings
- Detailed description
- Claims
- Abstract
- Drawings

A full example of a utility patent, illustrating these parts, is found in Appendix C. Of all these parts, the claims are the most critical to a good patent, so we'll cover the others briefly and then dig into the claims.

Background

The *background* section of the application explains the field of the invention and the problem your idea solves. When well written, it establishes the problems people face in the field and describes how they have not yet been solved by other inventions. At its best, the background begins to establish for the reader that there is a compelling need for the invention that is about to be described.

Summary

The *summary* is a brief description that begins to tell the story of how your invention solves the problems in the field.

Brief Description of Drawings

The *brief description of drawings* section lists the drawings that are included in the application and gives a short summary of what each one contains.

Detailed Description

The *detailed description*, sometimes called the "detailed description of preferred embodiments," is the bulk of the patent application. The detailed description is a narrative that tells the story of your invention. It describes what your invention is, what parts it includes, how those parts interact, and how it is used. It describes in words, very precisely, how your invention is made and how it works. When writing the description, discretion is used to determine how much detail to include regarding the "old part" of the invention. If your idea is an innovation (an improvement on an existing invention), you will mostly focus on the improved part and include some details of the prior art portions that are necessary to establish context for your improvement.

The detailed description will usually refer to the drawings throughout the discussion. Typically, the elements are numbered, and corresponding reference numerals are provided in the drawings. The specification might say: "Referring to FIG 6, the toothbrush 10 includes a handle 20 and a brush head 30. The brush head 30 has bristles 32." And if you look at FIG 2, you will see that those portions of the toothbrush are labeled with the same reference numerals. In most cases, this is how the detailed description is reviewed, by referring back and forth between the written parts and the drawing figures.

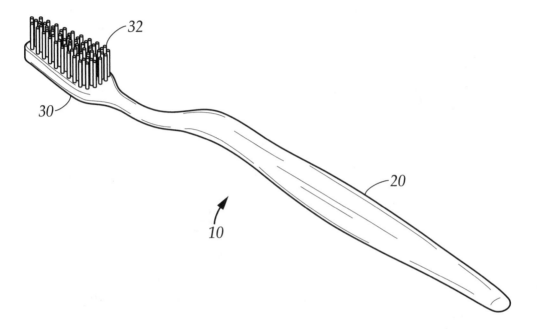

As we discussed in Chapter 2, your patent application must include an enabling disclosure, which is defined as enough information that anyone who is familiar with the subject area of your invention could make and use it without "undue experimentation." In other words, the application must provide enough detail that others can reproduce the invention without going to too much trouble or having to guess too much to figure out how to do it.

Consequently, you will often find details in the description and drawings that go beyond what actually makes the invention special. This is because part of what's needed in a patent application is to provide a workable example. The core concept might just be for a chair with four legs attached to a seat, but the drawings might show each leg attached to the seat with a four-way flange and screws that go through the flange into the seat. As we discussed earlier, this doesn't mean the patent is limited to the chair having that four-way flange, nor is it limited to having four screws. The added detail provides a workable example of how the legs *might* be attached to the seat.

Why would you need to give others the ability to reproduce the invention? Isn't the whole point of a patent to stop others from reproducing the invention? Yes, but the point of a patent is also that, once the patent expires (in twenty years for a utility patent), other people will be free to make it. In fact, the government gives you the protection of a patent *in exchange* for an explanation of your invention that's clear and detailed enough for others to produce the invention as well after the patent expires.

Drawings

While *patent drawings* are not technically necessary to have a complete patent application, they are required if they are "necessary for understanding the invention." As a practical matter, for nearly all inventions (except perhaps certain chemical inventions), drawings are necessary and must be included.

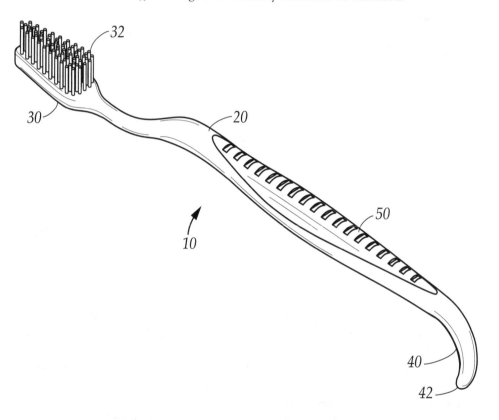

Patent drawings must follow the standards established by the Patent Office drawing rules. These rules dictate how drawings must be prepared and submitted to be acceptable. Numerous examples of patent drawings are provided throughout this book and also here in the figure below. As a general rule, patent drawings are black-and-white line drawings and must be easily reproducible as your patent application is processed within the Patent Office. Think: *what would copy well on a photocopy machine?* That's the way patent drawings need to be. So photographs as well as colored and shaded illustrations are not acceptable.

Abstract

The last section you will find in a patent application—the *abstract*—is limited to 150 words and summarizes the invention described in the patent application. The traditional purpose of the abstract is so that someone quickly thumbing through a pile of patents while doing a patent search can readily and quickly identify what your invention is about, so they would know whether they should study it in more detail.

Claims

Patent applications are written to get across what an invention is, how it is different from what has been done before, and why it is worthy of a patent. If you have a concept that is itself new, the intention is to patent the concept to the furthest extent possible. As we have noted previously, however, although the drawings and description may go into more detail and go further than just the concept, this does not mean that the protection is limited to what is shown in the drawings. *It's the claims that define the invention.*

If there is one thing you absolutely need to learn about patents before investing in one, it is patent claims and how they work. Before I explain that, let me give you an analogy to demonstrate why they matter:

Imagine there is a person (for simplicity's sake, we'll call him Bill) with whom you are considering working. Bill has represented himself as someone who is affluent and has significant resources to help you. You are trying to assess whether Bill is what he says he is.

Bill invites you to visit him at a beachfront apartment building. You pull up to the circular driveway in the front and are greeted by a valet who parks your car as well as a door attendant who ushers you into the marble and gold-decorated lobby.

You take an elevator to the forty-fifth floor, and the elevator opens up directly into the apartment. Bill greets you and brings you to a living room area surrounded by panoramic windows overlooking the ocean. If you had to guess, you would say this apartment must be worth ten million dollars.

There is no doubt you are very impressed by this apartment. You might immediately think that this guy is legit. He must be very wealthy to have a place like this. Or is he? Then you wonder: does he actually own this place? Maybe he just rents it. Maybe he is a guest of the actual owner. Maybe he just works for the owner. Or even, maybe, the real owner doesn't know he is using the apartment at all!

While it is a great and impressive apartment, you wonder: what exactly is Bill's ownership interest in it?

This is exactly the type of inquiry one needs to make about the claims of a patent to truly understand what that patent covers.

Consider that the disclosure in a patent or patent application is like the apartment Bill shows you; it can paint a very pretty picture of an invention, without really clarifying which parts are old and well known, and which are the new features that the patent actually covers. The disclosure doesn't tell the story of what the patent holder owns.

The patent claims section, however, is like the deed that Bill shows you to prove he owns the apartment. The patent claims define what is new in this patent—and what territory the patent holder actually owns.

Let's apply this to a scenario in which you encounter someone who owns a patent, and you wonder what she actually owns. Suppose this patent holder tells you she owns "the patent" for an electronic door lock that allows the user to open a door using her smartphone. Well, on the surface, that sounds like a very valuable patent! I mean, the idea of using your smartphone to open your door—that's a huge market, and it could be really big! But, the question is: does this person actually own *the* patent on *the idea of using a smartphone to open a door*? Maybe this patent is just for *certain features or a certain improvement* in a smartphone door-opening system? The actual truth about what this patent represents is what the patent claims will determine.

Once you understand how patent claims work, you will be able to see the difference between a patent on the concept and a patent on some minute details. You will be able to tell a patent that might be worth a lot from a patent that has little value.

Not understanding this, however, can cost you. It can lead you to seek a patent that won't be very valuable. It can cause you to believe that someone else's patent stands in your way, when in reality it doesn't. It might even cause you to believe you need to purchase a patent from someone before you can make your product, when you really don't need it.

Before we get into the mechanics of how patent claims work, let's look at how they come about, how they evolve, and how they are used.

The Lifecycle of Patent Claims

1. Patent claims are initially drafted and submitted as part of your utility patent application. In writing the claims, your attorney attempts to keep the claims broad by carefully choosing language that embraces what's new about your invention, without extraneous details.
2. The patent claims in the application are reviewed by the examiner and compared to the prior art. The examiner determines whether the claims "read on" the prior art—that is, whether they can be interpreted in a way that describes inventions that came before yours. If they seem to, the examiner may reject some or all of the claims in light of the prior art.
3. In a response, the patent claims may be revised by your attorney, in consideration of the Patent Office examiner's rejections.
4. Once approved, the issued patent will contain the patent claims as originally drafted, or with any amendments made prior to the approval by the examiner.
5. The patent claims in the issued patent are used by others to evaluate the "territory" owned by the patent, and whether they would be infringing your patent with their product.
6. If you bring an infringement action against an alleged infringer, it's your patent claims that will be considered by the court to determine whether the other party is truly infringing.

In the end, it is only the claims contained in the issued patent that really matter for establishing the value of the patent.

In light of this, understanding the general principals of how claims are structured and how they are interpreted will help you to:

- Make a good decision about whether to file a patent application.
- Know whether your competitor's patents might stand in the way of you launching your product.
- Make good decisions about whether to make changes to the claims in your patent application when trying to overcome a rejection from the Patent Office.

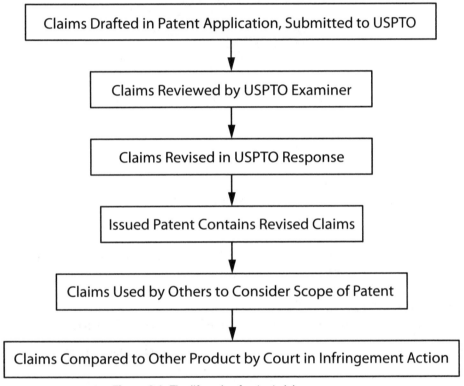

Figure 6.1 The lifecycle of patent claims

The interpretation of the patent claims in the prior art is critical to writing a strong patent, as well as in determining infringement.

How Claims are Interpreted

The claims recite *all the elements that another product must have to infringe the patent*. Each claim is a description of certain elements of your invention. Let's call the elements A, B, C, and D. For a product to infringe the claims, it must have every element recited in the claims. If the infringing product just has A, B, and D, it is not an infringer. If it has A, C, and D, it's not an infringer. If it has A, B, C, and D—now we're talking!

Clearly, then, the "name of the game" in drafting claims is to keep it as simple as possible. But you can only make it simpler if the simpler definition does not

Figure 6.2 A chair

describe things that already exist. So if A, B, and C exist in the prior art, you might need to draft the claim to A, B, C, and D to get past the prior art.

Let's look at some examples. Imagine you invented the chair shown in Figure 6.2. In a basic sense, the chair has four legs and a seat. A basic patent claim seeking to broadly protect the concept of the chair might look something like this:

1. A device for sitting down, for use upon a ground surface, comprising:

a seat having a top and a bottom; and

four legs attached to the bottom of the seat, for supporting the seat upon the ground surface.

This claim establishes what your invention is and defines the territory that others would need to copy to be an infringer of your patent. If they make a chair with four legs and a seat, as described, they would infringe. If they made it with a seat, but supported it with a single pole, they would not infringe. This is where every word in the claim becomes important. The key to writing claims is to keep the definition as simple as possible. But the definition you create when your claim is drafted will only be acceptable to the Patent Office examiner as long as it does not describe things that already exist! If the examiner finds prior art that shows a seat and four legs supporting the seat, the claim will be rejected. Then you might need to add something more specific to the claim to distinguish it from the prior art. In this case, you might add the seat, or the arms, back into claim 1 to make sure it is distinct from the prior art.

Independent and Dependent Claims

Most utility patents have more than one claim. The majority of patents have a listing of about ten to twenty claims, and some have many more than that. In any listing of claims, there will be at least one *independent claim* and generally several *dependent claims*.

The independent claim stands by itself; it contains a complete definition of the invention. It usually begins with "A" or "An." In the example of a full utility patent in Appendix C, claims 1, 7, and 15 are independent claims. A dependent claim refers back to another claim and adds additional features onto it. Dependent claims usually begin with "The." For example, adding on to claim 1 in the chair example above, you might have these additional claims:

> 2. The device for sitting down as recited in claim 1, further comprising a back panel extending upwardly from the seat.

> 3. The device for sitting down as recited in claim 2, further comprising a pair of arms, each arm extending forwardly from the back panel, parallel to the seat, and then curving downwardly toward the seat.

So now claims 2 and 3 are dependent claims that add additional detail to claim 1. Claim 2 adds more detail to the definition created in claim 1, and claim 3 adds even more details to that definition.

In Order to Infringe Your Patent, a Product Must Infringe at Least One Claim in the Patent

If a product meets the definition of even one claim in a patent (of course considering whatever other claims that claim references), then the product will infringe the patent. Okay, so if the product only needs to infringe one claim, why do we have so many? Because the other claims are backup. They give us the ability to hedge our bets. They give us the ability to try new approaches and not put all of our "eggs" in the "basket" of just one claim.

Having More Claims Increases the Possibility that the Examiner Will Find "Allowable Subject Matter" in Your Application

Often, when the examiner reviews the patent application, she might go through the claims and think:

> Claim 1 is too broad; I have prior art that fits the language of claim 1. Claim 1 should be rejected. Claim 2 adds a back panel onto claim 1. I have prior art that shows the back panel too, so claim 2 should be rejected. But now, claim 3 adds the pair of arms. The prior art doesn't seem to show arms like that in combination with the back panel and the rest of the device.

What happens next is that the examiner issues an Office Action rejecting claims 1 and 2, but "objecting to claim 3." The examiner then says those magic words that every patent attorney loves to hear: "Claim 3 would be allowable if rewritten to include all of the limitations of the base claim and any intervening claims." What this basically means is: amend the claims to put the back panel and the arms into claim 1. Claim 1 will then become allowable, and you'll get a patent!

It is so much better to get an Office Action in which you already know some of the claims will be approved than to get one in which all the claims are rejected!

Having More Claims Provides a Backup if the Patent Is Ever Litigated

As we will discuss in Chapter 15, in a *patent infringement lawsuit*, one of the main defense tactics by someone who is infringing your patent is to try to show that your patent is invalid. Often this is done by digging for prior art that the Patent Office didn't consider before they approved your patent application. The infringer would then try to show that this "newly found" prior art means your claims were too broad and are invalid. So here's where having several claims will help. If some of the claims are found to be invalid, as long as at least one infringed claim is left standing, the patent is still infringed! So if claims 1 and 2 are knocked out during litigation, as long as the defendant is infringing claim 3, that will be good enough to win the case.

What Goes into a Design Patent Application?

Since a design patent protects the ornamental appearance of an item, it makes sense that the thing that is most important in a design patent application is how we portray its appearance. And what is the best way to get across the appearance of an object? Through drawings, of course!

Because drawings are used to convey the appearance of your invention, the drawings are critically important. For the most part, in a design patent, the drawings are considered the disclosure. So most of the rules regarding how accurately the description is written in a utility patent application apply to the drawings in a design patent application. Among other things, the drawings must provide an *enabling disclosure*. The drawings must convey the shape of the object in a way that provides no ambiguity. To do this, every side of the item must be shown in the drawing figures. And the drawing figures must be perfectly consistent among the various views. Refer to Appendix D for an example of a full design patent.

The actual written part in a design patent application is pretty minimal. It's essentially just a brief description of the drawings and a single claim. And the claim itself just refers to the drawings. For example, "I claim a bicycle tire, substantially as shown and described."

Because the description is quite basic, writing that part of the application doesn't require much finesse. Where finesse *is* required is in the drawings. Professional judgment and experience are critical to knowing what to put in the drawings! The rule about how much detail to put in (as we discussed in the beginning of this chapter) is kind of opposite for design applications as compared to utility applications. Because the claim in a design patent is for the object just as it is shown in the drawings, as far as what is protected by the design patent, it literally *is what it is*!

When the drawings are prepared, sometimes unimportant details are left out. Other times, some features in the drawings are shown in dotted lines, which means they are "disclaimed," as seen in Figure 6.3. If you look at the design patents for

Figure 6.3 A design patent drawing for a bottle stopper, where portions have been disclaimed

many popular products, you will notice that several have portions that have been disclaimed. What this means is that, in determining whether a competitor is infringing the design patent, while considering the overall appearance of the design, the jury would not consider the parts shown in dotted lines.

Summary

A patent application is a critical document because once approved, it becomes the actual patent. A good utility patent application describes the invention with an abundance of detail in the main part—known as the specification—and carefully limits the description in the claims. For design patents, it's critically important that the drawings unambiguously show the shape of the invention from all sides.

The companion site has further information about how patent applications are prepared at www.patent-book.com/patent-application.

Losing the Right to Patent and Other Pitfalls to Avoid

In the previous chapters, we talked about what it takes for your idea to be eligible for a patent. Let's call the following discussion the "unless you do this . . ." chapter, because here we are going to talk about ways you can "mess it up," and lose out on getting a patent that might otherwise have been obtainable.

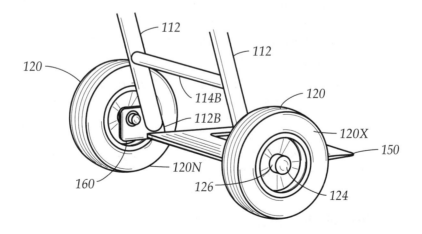

Disclosing Your Invention Publicly Before Filing May Result in Losing the Right to Patent

This pitfall needs to come first because, over the years, this one is probably the biggest reason I need to tell a client with an otherwise promising invention, "Sorry, I can't help you." Since the law has changed recently on this, and it has gotten much more complicated, and much more harsh; here's the simple fact: *It's a bad idea to publicly disclose your invention before you apply for a patent!*

Disclosing your invention before filing may kill your chances to get a patent in the United States and/or abroad. The safe bet is to file before you publicly disclose. Under some circumstances, you *may* have a "grace period" in the United States. Note that in many other countries, including Germany, the United Kingdom, and China, there is no grace period. In these so-called "absolute novelty" countries, if you publicly disclosed your invention before your first patent application filing, you are barred from ever getting a patent.

To clarify when you might still be able to get a patent in the United States, even after you publicly disclosed your invention, requires a more complicated explanation. Here are a couple of principles to chew on:

Principle Number One

> If your invention was in use or on sale within the United States more than a year before you apply for a patent, **it is too late.**

Plain and simple. No excuses, no getting around it, if more than a year before you apply for a patent, your invention was public in certain ways—it was on sale, or it was in use publicly (for example: displayed at a trade show)—it's too late to apply for a patent. They call this the "one-year rule" or the one-year "statutory *bar*." If it applies to you, it is said that you are "barred" (from getting a patent).

This is the way I've seen it play out, way too often, during my career:

  A. Someone has a product idea, and says, "If it does well, then I'll try to patent it."
  B. He or she starts selling the product, and two years go by.
  C. The inventor then comes to me and says, "I've just got to patent this! It's selling really well, and I know other people want to copy it."

That is the moment when I must give this entrepreneur the unfortunate news: *Since the product has been on sale for more than a year, it is too late to try to patent it.*

It's a really harsh rule. And it is so unexpected for most people! They often ask, "How would anyone even know that?" I've often thought, *if there's just one thing they should teach everyone in grade school about patents, to save them from unwittingly losing out someday when they invent something, it's about the one-year rule!*

Principle Number Two

> Under certain circumstances, if your invention is described in a publication anywhere in the world, you will be barred from getting a patent.

This means that you can never get a patent on the idea if it has already been disclosed publicly:

- **by you** (meaning you published it, or it was your product that was described in the publication)—unless you file your patent application within a year of the publication's date.
- **by someone else** (meaning it was a competitor's product described in the publication), and you have not already filed, *unless your invention was disclosed publicly earlier than theirs **and** less than a year ago.*

This is what they call the "one-year grace period." Ironically, in the latter case, your public disclosure would actually help you to avoid being barred by the other person's earlier publication.

Yes, this is complicated, and I've only scratched the surface. The law changed significantly just a couple of years ago and already there are court cases where companies are fighting over the details of these laws. Why? Because someone publicly disclosed an invention before filing a patent application. If he or she had simply filed the patent application first, there wouldn't be a fight over the intricacies of this rule!

The bottom line is, it's safest to file your patent application before the invention is made public in any way. If you don't, subject to the complicated rules I discussed earlier, you may lose the right to patent it!

Being the "Second to File" Won't Get You the Patent in Our First-to-File System

For a long time, people believed that as long as they could prove that they were the first to invent something, they didn't need to file a patent application. As long as they had proof that they invented it when they did, they believed they would prevail.

As a result, the myth of the "poor man's patent" continued. The underlying "theory" of the poor man's patent went like this:

Write down your idea, place it in a sealed envelope, and then mail it to yourself. The postmark will provide proof that you invented the idea. [DON'T ACTUALLY DO THIS!]

In the past, it took a lengthy explanation to convince people that the poor man's patent doesn't work, and why it won't be effective if someone else actually files a patent application and gets a patent on the same idea as yours.

But since the America Invents Act (AIA) was signed into law in 2011, the explanation for why mailing the idea to yourself doesn't work is a whole lot simpler: Under US patent law, the first to file a patent application is the one entitled to the patent!

You see, the poor man's patent myth was bolstered by the fact that, prior to the AIA, the United States had a "first-to-invent" patent system. In fact, we were the only country that did! Under the old system, in theory, the first person to invent could gain the patent over someone who invented later but filed a patent application first. However, since the AIA, which was implemented in part to align the United States with the rest of the world on this issue, we are now *first to file*. Under the first-to-file system, the first party to file a patent application wins, even if the other person can prove he or she actually invented it first.

There are a few limitations to the first-to-file system, however. The main one is called "derivation." That is, if the person who filed first did not actually invent the thing, but instead derived (copied) it from another, earlier invention, then the first filer would not be entitled to the patent. But since this is all brand new, little is known about how these derivation proceedings will play out. Most probably, this limitation will be useful in situations where a company insider steals research information and applies for a patent outside the company before the company can file its own patent application. But for most inventors, this exception won't matter. If a competitor files a patent application before you, it'll be hard to prove anything so sinister happened. The moral: *Get your patent application filed first to avoid losing the right to patent it!*

Disputes with Your Partners May Result in an Abandoned Patent Application

This one isn't exactly a way to lose the right to patent your invention, but it *is* a way I've seen stalemates happen where no one ends up winning. If you have one or more partners, and you're filing a patent application together, *don't have any disputes that jeopardize the partnership*. Just don't. Or work it out. Or at least figure out *who* the client is before you hire a patent attorney!

It's important to understand that, when you engage your patent attorney jointly with your partners, your patent attorney *cannot take sides* in your disputes. That would create a conflict of interest. If you hire the attorney jointly, the attorney conceivably has *every partner* as a client. If a dispute arises, the attorney can stop working with everybody involved or can keep working with everybody equally. The attorney can't "drop" any of the partners individually from the patent project.

This also means the attorney can't share information with less than the full set of partners, even at the request of one of them. Often when there is discord in the partnership, one might tell the attorney, "*Don't tell my partner this information about the patent,*" or "*Listen to my instructions, not his.*" In this situation, if one partner gives the attorney instructions, and there is doubt that he is speaking for all of them, the attorney will be unwilling to act unless certain that the whole partnership approves, and a stalemate begins. The attorney can no longer assume that communicating with one partner is as good as communicating with all of them.

What a mess! No one wants to be in this situation. It creates extra work, and it's distracting from the business at hand—getting the patent. Even worse, if you're being jointly represented by a patent attorney or law firm, a dispute between partners can quickly result in a conflict of interest that would require the attorney or firm to withdraw from representation! Then, if you want to continue pursuing your patent, you might all need to hire new attorneys! When it gets to that point, very often the situation is unworkable, and the patent application ends up being abandoned.

It's inevitable that entrepreneurs will disagree with each other—about a particular decision, or about overall direction and vision. Unless you want to suddenly have to hire a new law firm to continue pursuing your patent, you need to maintain a united front—at least as far as the business of obtaining a patent goes. The best advice is to keep your arguments and disagreements away from your patent attorney, so he or she can continue to represent your aligned interests in getting the patent. Or better yet, make an agreement ahead of time: agree with your investor in advance that the investor might be footing the bill, but *you alone* are the patent attorney's client; agree with your partners that only one of you will speak for all, and together let the attorney know.

Not Including the Correct Inventors Will Not Get You a Valid Patent

It's surprising how many people try to get around the inventorship rules for one reason or another. Some frequently heard excuses are:

- "Yes, I came up with the idea. But I want to put this in my wife's name."
- "I had the original idea, but my friend sketched it out and added this cool part, which I want to patent."
- "I came up with the idea by myself, but my partner is putting up the money. So we want to put both of our names on the patent."
- "I'm the owner of the company; one of my engineers had this idea. I want to patent it in my name."
- "We've been developing an app. One of our developers came up with a cool new feature, and we want to patent it."

All of these scenarios indicate a potential *inventorship* problem. Under the US patent laws, a patent must be applied for in the names of all of the inventors. Even when major corporations file patent applications, the inventors listed are the employees who came up with the idea. Typically, then, the listed inventors assign their rights to the company. Nevertheless, the names of the actual inventors are listed on the patent.

The same applies for innovations that happen on a much smaller scale. The actual people who contributed to the *claimed invention* must apply for the patent. Any attempt to circumvent this rule can jeopardize the validity of the patent. If you have any doubt about who contributed to the claimed invention and must be named, discuss it with your attorney.

While the patent must be applied for in the names of the inventors, the patent itself (and the patent application) can be owned by anyone. An assignment document can be filed with the Patent Office to transfer ownership to another individual or another legal entity (such as a corporation, limited partnership, etc.). A common practice when there are investors is to assign the patent application to a corporation; the ownership of the corporation is divided among the inventors and the investors in a manner agreed to by all.

If you are a company owner or an entrepreneur, and you have other people develop things that you might want to patent, consider that there may be a future inventorship issue. Perhaps then you can appreciate why it's important to have agreements already in place—to make certain that they will sign the patent applications and assignment documents when you request that they do so!

Not Reporting All Known Prior Art to the Patent Office Can Result in an Invalid Patent

As we discussed in Chapter 3, a person applying for a patent (and certain other people working with that person, including his or her attorney) has a duty to disclose to the Patent Office any information that the examiner might consider relevant when deciding whether to approve the patent application. This duty exists under the patent laws and rules—and you specifically take on this duty when you sign the patent application declaration. This duty continues even after your patent application has been approved, *up to the day that the patent formally issues*.

Failing to disclose key prior art violates your duty of disclosure as a patent applicant. It can be considered *fraud on the Patent Office* and can be used against you later, in patent litigation, for a finding of inequitable conduct and/or to invalidate your patent.

Not Responding to the Patent Office in the Time Allowed Can End Your Chances

This should go without saying, but I'll say it anyway: If you miss a deadline or fail to pay a fee on time, the application will be abandoned, and you can lose your rights to the invention.

Employment Agreement Limitations May Prevent You from Filing

It's a fairly common practice for employers of all kinds to have their employees sign an employment agreement. Many of these agreements have an *intellectual property clause* indicating that certain things you create will belong to the company. What this clause says varies from contract to contract. Some of them simply state that things you create related to your work with the company belong to the company. Others, however, say that virtually *anything* you create that's worthy of a patent, trademark, copyright, or trade secret, belongs to the company!

These clauses also vary in terms of penalties for not complying. For example: In some cases, if the company later discovers that you have patented something, the company might claim ownership of the patent. Under other contracts, not disclosing to the company that you have invented something can cost you your job—or even subject you to a penalty.

Because your employment agreement might affect whether you pursue a patent, take the time to dig it out (or request a copy from your human resources department) and show it to your attorney. If the agreement does seem to stand in the way of you pursuing your invention, you can often ask your employer for a letter granting an exception so that you may proceed. Whether he or she will give you this exception is more about the strength of your relationship with your employer than anything else.

We've been talking about dealing with the employment agreement you might have already signed. But it also pays to be proactive and have this in mind the next time someone asks you to sign one. If you switch employers, and you are asked to sign an employment agreement, keep an eye out for an intellectual property clause and negotiate if necessary. Especially if you are inventive, it pays to clear the way for any side project that might come about during your employment. You might even tell the employer that you are creative, or that you are already working on an invention, and you want to exclude that from the agreement. It's not unreasonable to ask for an exclusion for things you create outside of work and outside of the scope of your employment. Of course, the employer might say no, but there's also a decent chance that he or she will say yes.

Summary

The most common way people lose the rights to their invention—often without realizing until it's too late—is to publicly disclose the invention before they file their patent application. Other pitfalls to avoid include employment agreements, partnership disputes, and being too late to file because a competitor beat you to it.

Should You File a Provisional Patent Application First?

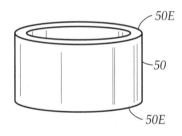

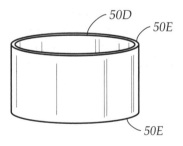

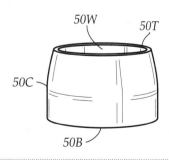

As we discussed in Chapter 4, a provisional patent application can be filed as a stepping stone toward a utility patent. A well-written provisional will establish priority for you, just the same as if you had filed a utility patent application. You'll keep this priority as long as you file a utility patent application that refers to the provisional within one year of the provisional filing date. Once you file a provisional, you can legitimately tell others that your invention is "Patent Pending"—just the same as if you had filed a utility patent application.

What Are Some Advantages of Filing a Provisional Application First?

Provisional applications do provide some strategic advantages in the patent process. They can help you:

- To invest less, yet obtain patent pending status
- To buy time before deciding whether to invest in a full utility patent application
- To "get you in the game," and stop the clock on potential public disclosure issues (and other pitfalls discussed in Chapter 7)
- To make sure that you are the "first to file"
- To give you some "lead time," when you know you have a solid concept but you also know your invention will undergo much development and many changes in the coming months
- To slow down the process, when it pays to delay having your application actually be examined at the Patent Office

Accordingly, there are definitely some advantages to filing a provisional that could benefit your situation.

What Disadvantages Are There to Filing a Provisional Application First?

The downside of filing a provisional is that it may put you at a slight disadvantage because:

- Once a utility patent application is filed, it will take some time before it will be examined, approved, and granted as a patent. If you file a provisional first, however, you are slowing the path toward having the issued patent in hand—by the amount of time that you wait before filing that follow-up utility patent application.

- A provisional is *sometimes* considered less seriously by investors and potential venture partners than if you had filed a full utility patent application.

If you want to take the extra time, and your investors understand the strategic advantage, it won't hurt your chances of getting a patent. But be aware that it might hurt your chances of getting funding to further develop your idea. It all depends on the people you're dealing with.

What's In, What's Out, and Why

When the provisional patent application was first created in 1995, the intention was to allow people to establish priority with a somewhat abbreviated, less formal application. But it still must be mostly written like a utility patent application to effectively establish priority. The same requirements for a written description in a utility patent application—including providing an enabling disclosure—apply equally to provisional patent applications.

The application must contain enough information that the invention is described in the right way to effectively "drop anchor" at the Patent Office—to firmly establish that you've come up with this idea. Should you decide to file a full utility patent application, that application must reference the provisional, and the extent to which the utility application describes the same idea as was filed in the provisional determines the strength of the priority date claim from the provisional.

Technically, a full set of patent claims is not required in a provisional patent application. Not preparing a set of claims definitely saves time. Not needing to write claims means that you can defer some decisions about which combination of features you are going to claim. This is especially the case when you have an invention that includes several possible *points of novelty*. Basically, you can disclose all of it and, to an extent, decide later on which point of novelty you will focus your utility patent application.

Written Description

Your provisional patent application must contain a robust description to be effective. The description should discuss all of the various parts/components of the invention and how they interact. As with a utility patent application, "connecting the dots" between all components is absolutely essential.

Drawings

While non-conforming drawings won't typically cause a problem at the provisional application stage, it pays to make drawings that are generally up to the standards for patent drawings for the sake of their future use, assuming you continue the process later with a utility patent application filing. And while the

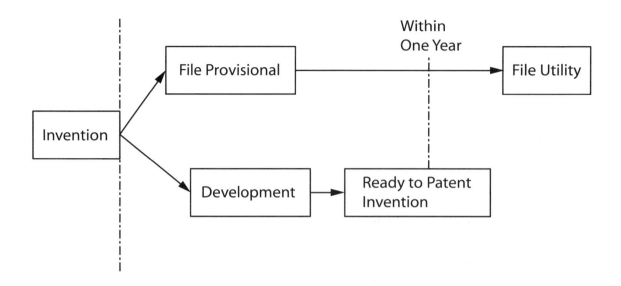

After a provisional patent application is filed, development continues and improvements may be incorporated into the utility patent application

totality of the rules regarding patent drawings are rather cumbersome to comply with, it's not difficult to make drawings that are generally suitable—at least for the provisional filing.

As we discussed in the previous chapter, and as seen in the examples throughout this book, patent drawings are black-and-white line drawings that pictorially depict the invention. Patent drawings are not like engineering drawings or blueprints, in that dimensions and projection lines are not included. If you use your provisional as priority for a US utility patent application, or even for a foreign patent application, the drawings with your provisional will become part of the file. If the drawings are not clearly reproducible, that might cause problems later.

What Is Necessary to Obtain a Filing Date?

While the substance of a provisional application is not reviewed by the Patent Office, it is checked for some basic items. If those items are present, a filing date is granted and a filing receipt will be sent to you. To obtain a filing date, a provisional patent application must include the names of all the inventors, their addresses, the address where any correspondence about the application should be sent, and the title of the invention. The application must also include a specification section that describes the invention, but even the most basic description will get you a filing date and filing receipt.

Another cautionary note: having these items will just ensure that you get a filing date. Whether that filing date will actually get you priority is a function of the completeness of the description that you provide.

The Potential Value of a Provisional

There are many ways to use the provisional application to your benefit. One effective strategy is to establish your priority at the Patent Office at a lower cost than that of a full utility application so that you can go out with the idea, talk to people about it, and see if it's worth the investment to file the full utility application. You can also further develop your idea before filing the utility application, and incorporate a more "ready" invention in the utility patent application. However, it's important to remember that a provisional patent application is only effective if a utility patent application is filed within the year, and a provisional is only as good as the writing within it.

Draft-It-Yourself or Drafted by Your Attorney?

This is where I get a little less adamant about my usual admonition that you must have a registered patent attorney write your application. While, to me, it is insane to try to write a *utility* patent application yourself—for a *provisional*, when time or funds are scarce, it may be a closer call.

Remembering that a patent application is only as good as its quality of writing, perhaps there are times when it's not as important that it be good. Sometimes the provisional application is being filed as a quick, temporary measure. Sometimes there isn't a lot riding on it. Maybe it was an idea that you came up with, and you wanted to get something filed quickly before you talked with anyone about it. In such cases, writing up and filing a quick provisional on your own might not be the worst way to go.

If you are going to write something up and file it as a provisional, at the very least, don't rely on it as you would one written by a patent attorney! Don't wait the full year before following it up with another filing, under the illusion that you are protected. In my opinion, the biggest problem with self-written provisional applications is the false sense of security they give to individuals. Inventors file their provisional applications, get a filing receipt, and then believe that they are protected. And since their provisional is not examined by the Patent Office, they won't get any feedback on whether the application is good enough to establish their priority. In fact, they won't know if it was good enough until it is too late!

One way of looking at the provisional is that, when you file your utility, you are basically saying, "I know I am filing this utility today, but look at my provisional—I actually invented this almost a year ago!" Thus, there are limits to how far afield you can go in the full application and still maintain the protection of the earlier filing date from the provisional. Clearly, the less the invention that is disclosed in your utility resembles the one that you disclosed in your provisional, the less effective the provisional will be. This is yet another reason to ensure that your provisional application is written almost as if it were a utility patent application!

Filing Your Provisional

After you file a provisional application, the Patent Office will generate a filing receipt that will be mailed to you. The filing receipt will show your filing date. Remember, receiving this filing receipt is the last you will hear from the Patent Office about your provisional application! As we discussed, provisional applications are not reviewed, so it's up to you to take any further steps. In general, the next step would be to file a utility patent application, claiming priority from the provisional application.

With certain limitations (that you should discuss with your patent attorney and consider carefully), you may be able to file a provisional patent application up to a year after you've made a public disclosure of your invention. It's important to note, however, that if you are considering filing for a foreign patent, such a public disclosure may disqualify your idea from patent protection in other countries. Also, if there *was* a public disclosure, and you don't file a full US utility patent application within twelve months of filing the provisional, you will lose the right to patent the idea in the United States *forever*.

A provisional patent application must be well written to effectively establish priority!

Allow Adequate Time to Start the Utility Application before the Filing Deadline

The deadline for filing a utility patent application is exactly one year from the filing date of the provisional. If you miss this deadline by even a single day, you will not be able to claim priority from your provisional, and the deadline cannot be extended.

If you are working with an attorney, be sure to provide adequate notice before the deadline that you would like to file a utility patent application. Most patent attorneys schedule their work several months in advance, so your best bet would be to contact him or her at least a few months before the deadline to discuss filing the utility. At that point, you should provide details of any improvements or changes that have taken place since the provisional filing. Also, if this attorney did not prepare your provisional patent application, allow even more time for him or her to get your provisional into shape for filing as a utility patent application.

Summary

A provisional patent application is a somewhat abbreviated application, but it must contain enough information that the invention is described in the right way to effectively "drop anchor" at the Patent Office—to firmly establish that you've come up with this idea. It can be an effective step to establish your priority while you further develop your concept and decide whether to file a utility patent application. To maintain priority from the provisional application, your utility application must be filed within one year of the provisional filing date.

There are many ways to use the provisional patent for your benefit. One effective strategy is to establish your priority at the Patent Office at a lower cost than that of a full utility application. You can then go out with the idea, talk to people about it, and see if it's worth the investment to file the full utility patent application. This book's companion site has further information about effective use of provisional patent applications at www.patent-book.com/provisional.

After the Application Is Filed

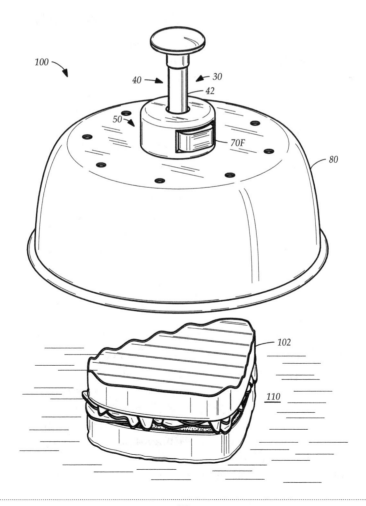

People commonly ask, "When I mark my product as 'Patent Pending,' should I include the *serial number* of my patent application?"

The answer is no. It's not required that you mark your product with the application serial number. Just the phrase "Patent Pending" will do and is preferable. When competitors see "Patent Pending" on a product, they might wonder (but can't usually find out) when your application was filed. However, because serial numbers are assigned sequentially, if you were to tell others your serial number, they would be able to estimate when you filed your application. Thus, telling others the serial number of your application gives up some of the "mystique" and advantage that normally surrounds your patent pending status.

So don't mark your product with the serial number, and others won't know how long ago you filed! Note that once the patent is approved and issued, it is *then* appropriate to mark it with the patent number. The difference is, when you actually have a patent, it will then already be public and is already readily searchable—so they can find out, regardless of whether you gave them your patent number.

CAUTION

Filing your first patent application in the United States Patent and Trademark Office can be very exciting! At this point, most inventors are filled with a sense of possibility and anticipation. In this chapter, we will discuss what your filing means, what to expect, and some things you may want to consider after your patent application is filed.

In sum, after your application is filed:

• Your idea will have *patent pending* status.
• You can expect a response from the Patent Office (if you filed a utility or design application).
• You should consider whether to file for patents in foreign countries and whether to file additional patent applications in the United States.

Patent Pending Status

One of the most significant milestones achieved by filing your application is that you will have established your priority for the subject matter contained in your application. Also significant: from this point on, it's appropriate to refer to your product as "patent pending," and the product and/or its packaging should be marked "Patent Pending." Your advertisements, website, and other literature should also notify others that your product is patent pending. Marking your product as patent pending puts others on notice that you are pursuing your patent rights. This "Patent Pending" notice can and should remain until your application is no longer pending, typically because either the patent has been granted or the application has been abandoned.

Patent pending status can provide a certain psychological protection against copying. When competitors see that your product is patent pending, they will of course want to know the details of what your patent covers. In most cases, assuming your application was filed with a *non-publication request* (see "foreign patenting" below), your patent application will be confidential, and your competitors cannot find out anything about it. They will not be able to assess what you are claiming or gain much certainty about the prospects of your patent being granted or what it might cover. This uncertainty can act as a significant deterrent from copying. If they can't determine whether they will be infringing whatever patent you might be granted, or when it might be granted, it's hard for them to assess the risk of copying you. Put differently, it would be a rather bold move on their part to start copying you, because there is no evidence that it would be safe to do so.

It's important to understand that you cannot stop someone from making, using, or selling your invention while the patent is pending. Consider that this is also where the psychological protection of patent pending status comes into play. Competitors know they would already be taking a risk in copying you by the fact that they can't determine what your patent might/will cover. But they take an even bigger risk because they cannot find out *when* you filed your application (unless you tell them). Even if you only just filed last month, for all they know you might have filed two years ago! It would then be an even bigger risk for them to tool-up now to manufacture and market your product idea, because your patent might be much farther along in the process and might conceivably issue next week or next month!

Expect a Response from the Patent Office

As we discussed in Chapter 5, after your utility or design application is filed, it will first be given a superficial review at the Patent Office, to see if it is complete and ready for substantive examination by an examiner. At some later point, your patent application will be assigned to an examiner, who will review it and either approve or reject it. A rejection comes in the form of an Office Action. This is an often lengthy document that details the rejection and provides a time period (typically three months) for your response. If the Patent Office does not receive a response from you within the time period allowed, your patent application will be abandoned.

Understanding the Different Types of Rejections in Your Office Action

The decision by you and your attorney as to how, or whether, to respond is usually a question of the types of rejections and objections made. Here is an overview of the types of rejections that might be contained in your Office Action.

"101s"

They are nicknamed "101s" because they fall under section 101 of the patent laws (35 U.S.C. § 101). A "101" is a rejection stating that the invention does not contain the proper subject matter for a patent. You would get a 101 if you tried to patent a song, because a song does not contain the proper subject matter for a patent—that is, it's not a machine, manufacture, composition of matter, or process. In recent years, however, 101s have become very popular rejections for patent applications covering computer software (see Appendix A) and business methods. Sometimes, you might even receive a 101 for "attempting to claim a human organism," if the wording of your claim includes the user of the product in the claim (e.g., the hand of the user swinging a golf club) in a way that the examiner finds improper.

"102s"

Again, they are nicknamed as "102s" after 35 U.S.C. § 102. This type of rejection is an assertion that there is a prior art reference that has the same invention as you have claimed. Note that the prior art reference cited by the examiner might not actually be the same invention, but the examiner thinks your claims are worded in a way that describes (or "reads on," in USPTO terminology) the prior art reference. In such cases, a 102 can likely be overcome by rewording the claims.

"103s"

Named after 35 U.S.C. § 103, this type of rejection is an assertion that your invention is obvious. Typically, this rejection is made based upon a hypothetical combination of two or more prior art references. In a 103, the examiner is not saying that one particular prior art reference has the same invention that you are claiming. Instead, the examiner is saying that *some combination* of prior art references amounts to your invention, *and* that it would be obvious to "a person of ordinary skill in the art" to combine those references. 103s can be tricky because examiners often take significant latitude in pulling different pieces together to reject your claims as obvious.

This is where experience plays a huge role in responding. You must have someone on your side who can evaluate the logic of the examiner's rejection and determine whether it is incorrect and can be overcome. Experience and strategy are critical here to come up with a plan for the best way to overcome a rejection—by arguing against the rejection, by amending the claims, or by some combination of the two.

"112s"

Named after 35 U.S.C. § 112, there are several different flavors of this type of rejection. In some situations, a 112 can be really bad, and in others, it can be easy to fix or overcome. But all 112s have to do with how accurately and how consistently the invention is explained in the specification and claims. Sometimes they are easy to overcome, when a misplaced word or two results in the rejection. Other times, they can be deadly. A 112 contending that the specification is "non-enabling"—that is, it doesn't provide enough detail—can be difficult to overcome. There is a rule that you can't add anything new—known as *new matter*—after the application is filed. Consequently, if the 112 rejection is because the examiner contends that a critical part of the claimed invention was not explained, leaving an insurmountable gap between the claimed elements, at this point you cannot add anything to the application to fill that gap! Your only option might be to argue that the examiner is incorrect in contending that such a gap exists.

A bad 112 rejection can sometimes be the reason to abandon an application, or to file a new patent application, known as a *continuation-in-part*, to add the omitted subject matter. Once again, knowledge and experience can help sort out the bad 112s from the no-big-deal 112s and come up with an appropriate strategy. In some situations, for example, there are tricks of the trade for skillfully fixing this type of problem. For example, if a critical part was not discussed in the specification (application text), but was clearly shown in the drawings, it might be possible to carefully insert a description into the specification to solve the enablement problem. This inserted description would not be considered new matter as long as this added text doesn't say anything that wasn't already crystal clear in the drawings originally filed.

Objections

An Office Action may also contain objections to the claims, the drawings, the title, or even the specification. Typically, these are minor matters such as a typo, misspelling, grammatical error, missing reference numeral in the drawings, the same part being given different names at different places in the application, and so on. Sometimes they are clearly errors, and sometimes they reflect a misunderstanding by the examiner. While it could be time consuming to deal with such objections, they are typically easy to overcome.

This summarizes the most common problems you will find in a Patent Office Action. The tricky thing can be figuring out which one is actually "controlling"— that is, which one makes it clear that you and your attorney need to pursue a certain strategy for response, that you should take some other action (like filing a continuation-in-part) or that a response would be pointless.

Consider Filing Other Patent Applications for the Same Product

Have you ever wondered why a product might be marked with more than one, or even several, patent numbers? It's because more than one patent application was filed. Sometimes, several patent applications are filed simultaneously. But more often, there was one initial application filed, known as a *parent application*, and then, sometime later, other applications were filed. These applications referred back to that parent application—and claimed priority from it. These subsequent applications (sometimes called "child applications") are known to the Patent Office as *continuation*, *continuation-in-part*, and *divisional* patent applications. Many companies do this strategically and file patent applications on every incremental improvements made to the initial product idea, which leads to them owning several patents on the same product.

Continuation

A continuation is a patent application that is filed after an initial patent application and that contains the same description (specification) and drawings as the initial (parent) patent application. However, it may contain new or different claims from the parent application. The continuation patent application maintains priority from the parent. A continuation must be filed while the parent application is still pending—that is, the continuation must be filed before the parent

IMPORTANT TIP

Keep in mind the possibility of filing continuation-in-part (CIP) applications while you wait for your first patent application to be reviewed!

After your patent application is filed, you will likely continue to develop your product. As it develops, it's quite likely that there will be improvements that may themselves be worthy of patenting. Continuation-in-part (CIP) applications, discussed also in Chapter 5, give you the ability to file applications on your improvements and enhancements, while maintaining priority for the main product idea as you had originally filed it.

This is a strategy that leads toward the possibility of creating a portfolio of patents surrounding your idea, instead of just a single patent. But remember that any new features do not have a priority date at the USPTO until an application describing them is filed. If you wait until your original patent application is examined before doing something to protect the new features, for various reasons—including public disclosure of those new features—it might be too late by then to protect them with a CIP. Thus, after your patent application is filed, it's important to keep this in mind as your product develops and as you wait to hear from the Patent Office regarding your original patent application.

application is abandoned or issued into a patent.

Continuation-in-Part

A *continuation-in-part* (CIP) application is like a continuation, except that, along with the old subject matter from the parent, it contains some new subject matter in the specification and/or the drawings. The CIP will maintain priority *only* for the portion of the subject matter that was contained in the parent; it will have a new priority date (the filing date of the CIP) for the new subject matter. As we mentioned earlier in this chapter, sometimes a continuation-in-part is filed when an Office Action shows that adding some new detail will be necessary before a patent will be approved. Other times, a continuation-in-part is used to protect innovations and new features that came about since the original patent application was filed.

When improvements are made after an initial application is filed, a CIP is often filed that includes the improvements. If the initial (parent) application is kept pending, both will be examined, and each can result in a separate patent being granted, as shown in FIG 9.1, below.

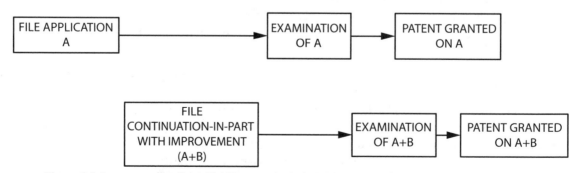

Figure 9.1 Improvements after initial filing may be included in a continuation-in-part patent application and possibly result in a second patent

Divisional

A *divisional* application is similar to a continuation in that it contains the same specification and drawings as the parent application. However, a divisional application is filed to pursue claims that were withdrawn because of a restriction. As discussed earlier, when a patent examiner contends that your application contains more than one invention, you are required to pick only one for the examiner to consider. The inventions that are not picked are considered "withdrawn from consideration." You may later file a divisional application to pursue the inventions that were previously withdrawn from consideration. Your divisional application must be filed while the parent application is still pending.

Post-Grant Review

There are various procedures for reviewing a patent after it has been granted, thus the name *post-grant review*. If someone thinks your patent shouldn't have been granted, that person can file a petition with the Patent Office to have your patent application given another review.

The first thing to know is that all of these procedures are relatively rare. A very small percentage of patents are ever involved in any type of post-grant review, so we won't go into detail about them here.

The procedures for reviewing a patent after it has been granted have evolved significantly over the last few years, and there are now both *ex parte* procedures and *inter partes* procedures for reviewing a granted patent.

The *ex parte* procedures take place as a dialog with the patent owner and the Patent Office. Just like how the process goes when you are seeking a patent, it's just you and the examiner.

Inter partes procedures, on the other hand, involve third parties. This means that a third party—another person or company—would be involved in the process of reviewing your granted patent. Such procedures are, then, more like a lawsuit in that there are two sides and a "judge." Obviously, the expense can rack up quickly. The thing to remember, once again, is that the chances of getting mixed up in any type of post-grant review are relatively small.

Foreign Patenting

There are three things I typically tell clients when they ask about foreign patents:

1. *It's very expensive!* Gaining patents in foreign countries means eventually filing a patent application in every country. Obviously, the expense quickly adds up, not just for filing all those applications, but also for pursuing each one through its own Office Actions and responses.

2. *It might not be as necessary as you think.* Your US Patent protects against anyone making, using, or selling the product in the United States. The frequent reason for wanting foreign patents—"*What if they make my invention overseas and try to sell it here?*"—does not warrant filing a foreign patent because selling it in the United States is already covered

IMPORTANT

Utility patent applications are often filed with a "non-publication request." If you didn't file this request, your patent application would be published eighteen months after your earliest priority date—whether or not a patent had been granted.

Since publishing the details of your patent application before you have the protection of a patent seems like a bad idea, you may want to request that the Patent Office not publish it. The condition is that in the non-publication request, you must assert that you *don't intend* to file foreign patent applications. And this is ok. It doesn't mean you cannot file foreign patent applications. It just means that *once you decide you will file foreign applications, you must rescind the non-publication request before you file foreign patent applications.*

This step is very important! So, if you do decide to file any foreign patent applications, and you use a different attorney to file them, make sure that your new attorney knows about your non-publication request! He or she will probably ask and will most likely request a full copy of your utility patent application from the attorney who filed it. But I figured telling you about this adds an extra safeguard to make sure it doesn't get missed!

by your US patent. Often, a US patent can give you sufficient leverage, since an illegitimate product made overseas cannot be sold here and technically cannot even be used here.

3. *You typically have a year from the filing of your first US patent application to begin pursuing foreign patents.* Now this part is key, since we're talking about what to do after your patent application is filed. You should make the decision about whether you want to pursue foreign patents at all *before* the one-year anniversary of your first filed application (whether provisional or utility). In *some* circumstances it may be possible to file for a foreign patent later, but don't count on it being possible if you miss the one-year date!

Maintaining Foreign Rights While Saving Money

Although the general rule is that you need to file in each country, there are a couple of ways to reduce or delay the expense of doing so:

- File a *Patent Cooperation Treaty* (PCT) application: A PCT application will not actually get you a patent in any particular country. What it does, however, is allow you to designate the countries where you expect to file, and then gives you up to thirty months to actually file those applications (see FIG 9.2, below). It also can centralize part of the searching and examination procedures so that when you do file in the designated countries, the examination process there will be smoother.
- File a *European Patent Office* (EPO) application: The general rule is that you must file in every country where you want patent protection. However, an EPO application gives you the ability to pursue patenting in the various countries of Europe as a single unit. There are additional costs, including

validation fees and annual maintenance fees, which can get costly, but it's certainly much cheaper than filing in all the individual countries of Europe.

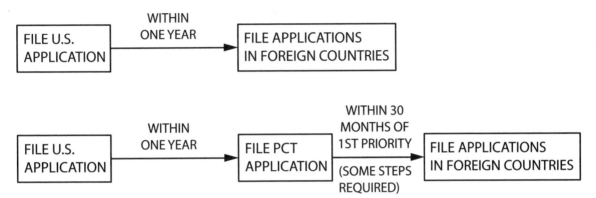

Figure 9.2 Filing a PCT application can provide additional time for filing foreign patent applications

Summary

After your patent application is filed, you will mostly be waiting to have it reviewed by the Patent Office. But you should be aware of various circumstances under which further action is necessary. If you make improvements to your invention, you might want to file additional patent applications to protect those improvements. If you want protection in foreign countries, you need to consider your options before it's too late, since the process generally must be started within one year of your first filed patent application (regardless of whether it was a utility or a provisional application that was filed first). For more information, visit the companion site for this book: www.patent-book.com.

CHAPTER

10

Working with a Patent Attorney

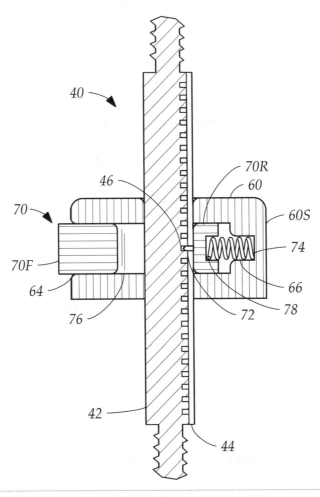

Confidentiality: Is It Safe to Talk to an Attorney About Your Idea?

Since this is a common question or concern that new clients have when they call my office, we might as well start here!

It is natural and understandable to have this concern/question. After all, your idea at this point is secret. You haven't told anyone. And the very fact that you even want to talk to a patent attorney means that you consider your idea to be valuable! So, is it safe?

The truth is—yes, it is safe; attorneys are bound to keep confidential whatever information you disclose to them about your invention. They can't tell others, and they can't use the information for any other purpose than to help you protect your rights.

Finding a Patent Attorney

The first thing to understand is that, in order to represent an inventor seeking a patent, an attorney must be a "registered patent attorney." A registered patent attorney is one who has been licensed by the USPTO after taking a specialized bar examination known as the "patent bar." To even be eligible to take the patent bar examination, the attorney must possess technical training—typically an undergraduate degree in engineering or science. Because of these requirements, overall a rather small percentage of attorneys are registered patent attorneys.

Because patent law is different from other areas of the law, it's not always easy to get a friend's recommendation for a patent attorney. For other areas—like wills, family law, or personal injury—you are likely to know someone who has used an attorney in that area of law. But with patents, unless you have a friend or colleague who previously pursued protection for an invention, you might not know anyone who has worked with a patent attorney!

As with the choice of almost any professional service, people tend to go with the first option that meets the criteria they consider most important. The problem is, if you have no experience hiring a patent attorney, you wouldn't know what criteria to use!

Considerations When Selecting a Patent Attorney

Some of the top criteria people use when picking a patent attorney are:

- Geography
- Technical background
- Cost
- Personal compatibility
- Communication style
- Available fee arrangements

The weight you give one or more of these criteria—or any others you may have—is for you alone to decide. The following discussion may be helpful in considering or discounting any of them.

Geography

Not having any experience picking a patent attorney, many first-time inventors naturally put "one who is nearby" at the top of their list. This stems partially from habit—after all, that's how we shop around for just about everything—partially from the fear of disclosing their invention to someone they have not met in person, and partially from the perception that it's important or necessary to sit face to face with the attorney so that he or she can understand the invention.

While this may seem logical, as a practical matter, it's not actually necessary that you sit face to face to work effectively with a patent attorney. With good communication, it's not essential for your patent attorney to be nearby. In fact, over the years, I have worked with literally thousands of inventors/clients whom I have never met. I even have some clients with whom I've worked for more than fifteen years, on several patent projects, and we've never met in person! It's also funny that many of my clients originally contacted me because I was nearby, but we ended up working over the phone and have never actually met.

Understandably, some people have a strong preference for working with people they have met face to face. At the same time, many people—especially people who work in the tech industry—are accustomed to never actually meeting the people they work with. Many of them have jobs where their co-workers all work remotely. To them, hiring someone they have never met in person seems perfectly natural. This is an example of where it is important to know yourself and what works best for you.

Technical Background

All patent attorneys who are registered to practice by the United States Patent and Trademark Office *must* have a technical background. For example, my technical background is electrical engineering. The types of technical backgrounds my colleagues have, however, can vary significantly. Some patent attorneys have engineering backgrounds, such as mechanical engineering or electrical engineering, while others have science backgrounds, such as chemistry, biology, or physics, and still others have computer science backgrounds.

Knowing the technical background of the patent attorney and whether it's a match for the technology area of your invention is a good starting point, but that shouldn't be the final basis for whether he or she is equipped to handle your project. I know many patent attorneys whose technical training was mechanical engineering, but they now mostly work on patents for computer-related inventions. In many cases, their experience counts for more than the technical training that others might possess.

In highly specific fields that involve and require significant advanced knowledge, you might truly need an attorney with an appropriate specialty technical background. For example, if your invention involves genetics, you would want someone with a technical background in biology or a similar field.

For most other types of inventions, however, it's not always necessary to have a direct match between the field of your invention and the attorney's technical background. Most experienced patent attorneys have worked on patent cases in a very wide range of technical fields. Being a patent attorney inherently requires you to be a "jack of all trades." Doing our job effectively requires us to have lots of general technical knowledge and an innate feel for "how things work." Whatever the field for the case we're working in, we dive deep enough into the technology to know how to explain and argue what is different about this invention from the prior art. So unless your invention is super technical and in a field that involves lots of specialized knowledge, technical background probably won't be your biggest consideration for finding an attorney to work with.

Cost

Understandably, I think that when people shop by price alone, it is mostly because they don't know what other criteria to use. All things being equal, why not shop by price, right? The problem is that all things are not equal. If you don't know how to differentiate them, however, then they sure seem to be!

Cost may in fact be an issue for you. If this is the case, I suggest doing the best you can to create a budget that's appropriate both for the value of your creation and for your finances. That being said, shopping by price alone is probably the worst way to pick a patent attorney. The work of all patent attorneys is not equivalent, so picking one simply because he or she is cheaper is a bad way to go.

Keeping that in mind, I do know there are quality patent attorneys who charge fees so very low that, quite frankly, they are underselling themselves! I know because years ago I was one of those attorneys. I was doing very good work for my clients, yet I charged too little. As a result, I was struggling to pay my expenses. But from the clients' perspective, they were getting great work for less money. Of course, it's not sustainable for a patent attorney to do this forever.

I can tell you from experience that there are many good patent attorneys who are shy about charging more (or even enough) for the quality of work they are doing. Sometimes these are younger attorneys who have worked at big firms and gotten great experience there and then have gone out on their own. Because they are newly on their own, they don't yet have many clients, and therefore they might charge less than they're worth.

If you search within your price range, you might find that diamond in the rough: the patent attorney who does a great job but charges too little. Let's face it; you can only afford what you can afford. But once you establish your budget or what you can afford to pay, I don't suggest using price alone to pick from among the attorneys in your budget range.

> **NOTE**
>
> An attorney does not need to be a "registered patent attorney" to represent you when you are seeking a *trademark registration* (see Chapter 16). Any attorney with a state bar license can technically represent you in trademark matters. It does pay, however, to work with an attorney who is experienced in trademark law.

Selecting an Attorney by Writing Ability

You might feel lost in trying to select the right patent attorney because our natural tendency is to want to hire someone who is good at what they do. When you try to pick a patent attorney that way, you've probably noticed that it is really difficult to judge how well any particular attorney knows their stuff!

But there is something that even as a layperson you probably can assess: their writing ability. For all the reasons we discussed earlier when we talked about drafting patent applications, your attorney's ability to write clearly is the most important skill they must have to not only help you get a patent, but also to help you get a good patent. They must possess the ability to write in a way that focuses on what's important and tells the story of your invention.

While writing ability is not necessarily as critical for attorneys in every area of the law, it is crucial if you are hiring an attorney that will potentially write your patent application. In fact, when I interview patent attorneys to potentially hire at my firm, to even be considered they must have excellent writing ability.

This may be the best "hack" yet for selecting a patent attorney! While you might not be able to tell which is the more knowledgeable patent attorney, you can probably tell who is the better writer. I'm not suggesting that you interview your potential attorney by asking for a writing sample, but often you can get a sense of how clearly they communicate by reading their correspondence or articles. This could give you a real insight into how well and effectively they will communicate your invention to the USPTO, and can provide you with the confidence that you have made the right choice.

Personal Compatibility

In life, some people "click," and some don't. Even in reading the previous sentence, some people know exactly what I mean, and some don't. In many cases, this is explainable by a difference in values and philosophy. But often, you can have an initial conversation with someone, and it's clear that you just don't look at things the same way. It pays to follow your intuition here. You don't need to force a relationship. Regardless of your values, philosophy, and style, you can probably find a patent attorney who thinks and acts more like you do and with whom you can relate better.

Communication Style

This is a factor that people rarely consider when picking professionals of any kind to work with. In my experience, however, this can be the single most important factor in having the relationship with your attorney go well. There are many categories to consider for this, but let's take a look at a couple of categories for communication style and how they can shape the relationship—or even create an immediate mismatch.

Responsiveness

There are some people you might call responsive. They usually respond to phone calls and emails within hours, and sometimes within minutes. There are other people who are—let's just say—*less* responsive. You might ask them a question and then wait days for an answer. And by the way, being less responsive doesn't mean less responsible. There are plenty of people who have a lot going on and do all their email and phone message responding once a day, or even twice a week.

Neither way is right or wrong, but put these two types in a working relationship together, and neither one will be happy with the other. The responsive person may think that the other is neglectful or even disrespectful. The less responsive person may think the other is pushy and demanding. One will be constantly waiting and trying to get the other to deliver, while the other will be constantly setting boundaries and trying to get the first one to back off and chill out. Immediate mismatch. Often, the clues to which type you are dealing with are in your first interactions. Once again, pay attention to this. If it's a mismatch, go elsewhere and save both of you a lot of wasted energy.

Listening, Relatedness, Empathy

How well your attorney listens to you, understands you, and really gets what matters to you can vary considerably. This is very important to some people, but not so much to others.

In this realm, here's another reality to consider: a patent attorney's job is probably 90 percent time spent with eyes glued to the computer screen, buried in a patent application, a response, or some other document, and 10 percent spent interacting with clients, colleagues, staff, and others. How the attorney spends that 10 percent, however, can make a big difference to his or her clients in whether it feels like their attorney is on track with their project.

For many people, those short interactions will shape their opinion of whether their attorney is taking care of them and even doing a good job. For others, they couldn't care less about the attorney's "bedside manner"; they base their judgments solely on the work product. Know which is your preference, and choose appropriately.

Patent Agents

In addition to patent attorneys, there are also professionals known as "patent agents" who can help you obtain a patent. Patent Agents are licensed by the USPTO to represent inventors seeking a patent and can conduct business with the Patent Office, just the same as a patent attorney. To be licensed as an agent, they must have a technical background and pass an exam, but do not need to also be a lawyer or have attended law school. Because they are not lawyers, patent agents cannot assist with drafting agreeements or handle any legal matters other than practice before the Patent Office in a patent matter. Since their practice is limited to just preparing and prosecuting patent applications, many patent agents are quite skilled in doing so. Generally it will be less expensive to hire a patent agent than a patent attorney.

Available Fee Arrangements

Finding out what types of fee arrangements the attorney is willing to use is a good start toward seeing if he or she is a match for you and your patent project.

Hourly

Probably the most common arrangement for attorney work is *hourly billing*. The theory behind hourly billing is: legal matters that require less time from the attorney should cost less than matters that require more time. When entering into an hourly billing arrangement, the attorney may give a time estimate. Such an estimate, however, is only an *estimate*. If preparing your patent application takes more time than expected, the bill will be higher than the estimate—sometimes significantly higher. As a modified hourly billing arrangement, some attorneys will agree to a *billing cap*. While the billing is still set by the hour, your agreement states that the total bill cannot exceed the cap.

Fixed Fee

As noted above, the majority of attorneys bill hourly for all the work they do. And for most bigger clients and corporations, this arrangement works out okay. But for smaller clients, the prospect of starting a project without knowing what it's going to cost is extremely intimidating! Luckily, these days more and more attorneys are using so-called "alternative fee arrangements." Among these fee arrangements is the "fixed fee." In the realm of patents, typically this means that the attorney will charge a flat fee for doing a patent search or a flat fee for drafting your patent application.

With regard to patents, a flat fee or fixed fee generally does *not* mean a single, flat fee for everything they will do to get (or attempt to get) a patent for you. Since the patent process is rather open-ended, it's hard—if not impossible—for an attorney to quote you a fixed fee for a whole patent project. The reason is that, at any twist in the road—such as a rejection from the Patent Office—there is usually a choice whether to do more work toward getting the patent or to drop it. Typically, this choice is made by considering what it would cost to proceed, along with the chances of success. Without an additional cost to proceed at each juncture, the client would naturally insist that the attorney exhaust every avenue, which would be prohibitively expensive if the client were paying at an hourly rate! So, clearly, it's not something the attorney can do under a fixed fee.

With that explained, then, it's easy to see why most attorneys who use fixed fees charge a new fixed fee at each step. My own practice is to propose a new fixed fee for proceeding at each juncture, and give the client the opportunity to choose whether to continue. I believe that many attorneys who charge fixed fees use a similar practice.

With a fixed-fee arrangement, the cost of the service is determined and agreed upon before the work is performed. Smaller clients often prefer this type of arrangement because it gives them the chance to fully understand what the final cost will be before authorizing the attorney to begin.

How attorneys determine fixed fees is totally up to their own sense of what the case is worth. Typically, this means using their own past experiences as a guide in making an assessment of how much time and attention the case will likely occupy.

A good fixed-fee arrangement is one in which it's clear what's included and what isn't. For example: In a fixed-fee arrangement for drafting a patent application, it should be clear whether the cost of drawings is included in the quoted fee. There should also be a clear process for determining whether revisions are included, and under what circumstances. For example, it would be reasonable for the stated arrangement to exclude revisions after the application has been drafted, should the inventor wish to add new features to the invention after the application is already completed. For the sake of the attorney-client relationship, this type of thing should be explained in writing—and understood by both parties—before the work begins.

Contingency Agreement

Some inventors hope to save their cash by offering a percentage of their invention in exchange for the legal services necessary to patent it. This type of arrangement is very rare, and while it may be referred to as a "contingency agreement," it's not actually a contingency fee.

A true contingency fee, as the name implies, is *contingent on the outcome*. For example: In a lawsuit, the lawyer is hired to obtain a certain result or outcome through providing his or her services. If the desired outcome is winning damages or a settlement in a court case, and the attorney is successful in achieving this, he or she would take part of the settlement as the fee.

With an invention, however, the outcome the attorney seeks to achieve is obtaining the patent. Making money from the patent is a separate deal—usually outside the control of the attorney and often occurring long after the attorney's work is done. So this arrangement is not actually a contingency fee.

Additionally, any arrangement in which the attorney acquires part of the property of the client is frowned upon by attorney ethics rules. It's not prohibited *per se*, but it requires that the attorney make specific disclosures to the client and explain the potential conflicts and other pitfalls that may arise from the arrangement, as well as the ways in which a more traditional arrangement might be in the client's better interest. Therefore, most attorneys don't consider it worthwhile to get involved in such an arrangement.

Another potential pitfall of this type of arrangement is that, in practice, it can be difficult for the inventor and the patent attorney to agree on what would be a fair percentage for the attorney to receive in payment. From the inventor's perspective, the idea is already very valuable, so the attorney should be willing to do the work for a very low percentage of this assumed value. From the attorney's perspective, the value of any invention is *highly speculative*. Regardless of how good the idea is, there are many factors that go into determining whether it will ever make a single dollar. So, compared to what the inventor is thinking would be fair compensation in such an arrangement, the attorney would typically be thinking that a very different, much higher percentage would be necessary to make it worth his or her while.

Paying Up Front

Especially in fixed-fee billing arrangements, many attorneys will require payment up front.

The advantage is that payment up front shows full commitment, which is great for the relationship. When someone pays up front, there's no doubt in the attorney's mind that the client is committed to proceeding. Some clients might truly *want* to proceed, but that's not the same thing as *being committed*. Actual payment is also a great sign that you can afford it and that you are not overcommitting yourself.

If an attorney is not asking for payment up front, this is not necessarily as beneficial to you as you might think. Unless there is an already existing relationship—and a basis for the attorney to trust your commitment—consider that there are drawbacks you may experience as a client for not having "gotten the payment thing out of the way." It can be hard for your attorney to go full steam ahead with your patent application when some part of his or her mind is wondering whether you'll be able to pay for it when it's done. When payment has already been made, however, wondering if you're going to pay won't occupy even a moment of your attorney's attention.

Retainer Agreements

Typically, when you work with an attorney, there will be a written retainer agreement. In many states, they are actually required by law. What types of things are addressed in a retainer agreement can vary considerably.

Here's the thing: a lot of clients don't read the retainer agreement. We are so accustomed to signing things put in front of us that we don't always read them. Who reads the "terms and conditions" before buying something or signing up for an account online? Hardly anyone. Most of the time, we only read something if we have reason to distrust the person asking us to sign it!

My suggestion is, regardless of how well you trust the attorney, read the retainer agreement. It's not just about trust; it's about creating a relationship. A well-written retainer agreement will set the context for the relationship, inform you about the process you are entering, and set your expectations. So even if you are in a rush and don't read every word at the moment you are signing it, make a point of reading it later.

Who Is the Client?

An attorney has a duty of loyalty to the client. The advice given and the actions taken by the attorney *must* be in the best interests of the client. Generally, whoever the client is, that is the person with whom the attorney will directly communicate and from whom the attorney will take instructions. When you have multiple people involved with your venture, however, such as inventors, investors, and other partners, it can become confusing as to who is the actual client. Is it the inventor? Is it the investor? Is it the inventor and investor together? Is it the corporation you have formed together?

This might seem less important in the beginning, when everyone is getting along and everyone seems to want the same thing. Later on, however, when there is a dispute or when the interests of the people involved start to diverge, the attorney needs to know whose interests he or she represents. We talked about this in Chapter 7—"Disputes with your partners may result in an abandoned patent application."

As a result, it's important to consider who the client is. Of course, it's the attorney's obligation to understand and clarify this where necessary. But you should understand this as well, because, if you have partners or investors, you might want to discuss this with them before approaching a patent attorney. In the discussion, for example, you might establish that, although they may be contributing money that will be used for the patent process, you personally will be the client of the patent attorney. Or, perhaps you might decide that the corporation you are founding together will be the client.

The Big Four "Don'ts" That Could Derail Your Relationship with Your Patent Attorney

Okay, I'd rather not even go into this territory, but I think it would be beneficial for us to talk about it. Maybe I'm just lucky, but I've had really great relationships with my clients over the years. However, I've seen other attorney-client relationships go off the rails. And typically, it is not because any one of them meant to cause problems; rather, it's more likely because they didn't realize the impact of their actions. I hope the following discussion serves to make you aware of the types of things that can be harmful.

1. Don't Suggest Doing Something Unethical or Illegal

It's important to understand that the actions you take with your patent attorney may have repercussions with your patent application, as well as with your working relationship. I probably don't even need to say this, but please don't ask your attorney to do anything illegal, unethical, or in violation of the Patent Office rules. For example, an inventor might say, "I invented this, but I'm going through a divorce, so I want to put the whole thing in my friend's name." Every patent attorney has heard this one. Most attorneys have no problem saying "no" to this request and giving solid reasons why not. But just making a request like this causes the attorney to trust you less—and wonder what other way you might ask him or her to skirt the law.

Here's a sample conversation regarding an unethical request. Patent attorneys hear things like this all the time:

Inventor: I found this patent for an invention just like the one we recently filed an application for.

Attorney: It's not exact; there are differences we could argue. But the Examiner for your application would want to know about this, since it is relevant to the patentability of your invention. We have a duty to disclose this patent to the Patent Office.

Inventor: In that case, can we pretend we never had this conversation?

Don't do this! When an attorney gets a sense that you are going to ask him to break the law—or even the rules—he won't want to represent you!

2. Don't Work with an Attorney Whom You Find Difficult to Fully Trust

If in the first meeting you question the advice and direction the attorney is giving, you might not be a match to work together. If what he or she is saying just doesn't sit right with you, trust your instincts and go elsewhere.

It may not be that what the attorney is saying is actually incorrect, but the alternative—proceeding with an attorney you don't fully trust—creates a poor working relationship. Consider that when you show doubt—especially about basic and common advice—he or she might make a judgment that *you* are the one who is difficult to work with! If you did continue the patent project together, from your perspective, you would never fully trust the direction being followed. And from his perspective, he will see you as a difficult client. Once again, probably not a good match—better to find someone whose advice and direction immediately make sense to you and feel congruent with your own approach.

3. Don't Be Disorganized with Your Information

While the fees that patent attorneys charge can seem high, the reality is that many I know find it hard to make a living. The ones who do realize that it's important to streamline their processes, so that they can give each case the attention it deserves while avoiding spending their time on unnecessary parts of the process. What this means is, they want to efficiently receive information about your invention so that they can use their time doing what they do best—shaping it into a patent application.

Patent attorneys are therefore sensitive to things that would create extra work for them. They are also looking for clients who fully "own" their project; clients who recognize that the patent attorney is there to do some specific things that require the attorney's unique skill set, while the clients are willing to do what's required on their end to get things ready—so the patent attorney can do his or her thing. Here are some examples of requests that might seem innocuous, but add a difficulty that the attorney will not appreciate:

Patent attorney: "Can you send me a description and some sketches?"

Client: "Just go to my website; it's all there."

Patent attorney: "How do tires like this usually work?"

Client: "If you Google 'bicycle tires,' you can find out."

Patent attorney: "Can you send me some photos of how this tool works?"

Client: "If you come by my shop, I can show it to you, and you can take pictures."

Attorney: "What's the purpose of this component of your invention?"

Client: "Here, I'll give you the number of my prototype builder. You can call her, and she'll explain it to you."

Once again, none of these things seems like a very big imposition. But to the patent attorney, who is trying to work efficiently, you just threw a curveball at his or her normal process of taking the information the client provides and shaping it into a patent application. The moral: keep it simple and respect the attorney's work process; the attorney will then see your case as one that fits nicely into his or her schedule.

4. Don't Give the Impression That You Are Unreliable

We all deal with busy schedules. And even the most reliable people sometimes are unreliable. When either the attorney or the client doesn't show up on time, doesn't return phone calls, or doesn't send back paperwork when they say they will, it makes it more difficult and time consuming to work together. The bottom line is that both should diligently avoid behavior like this, because inevitably, it harms the relationship and makes getting the work done much more difficult.

Summary

Finding and selecting a patent attorney to work with can be difficult, because it's hard to even know what criteria to use in selecting one! Find an engagement that works for you—not just in billing arrangements and cost, but also in whether you have a sense that it is a good fit to work with him or her. And be diligent in your preparation to help maintain a good working relationship.

For more information about working effectively with a patent attorney, visit the companion site at www.patent-book/patent-attorney.

What Is It Going to Cost?

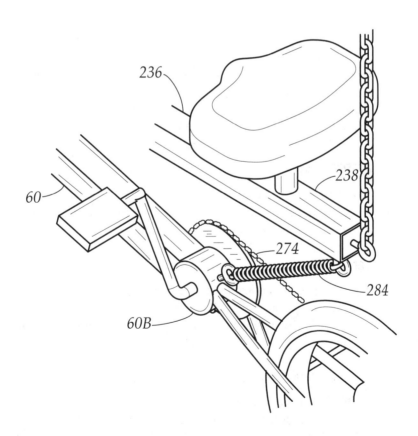

The cost of preparing and filing a patent application in the United States has several components: attorney fees, the cost of drawings (if billed separately), the cost of the prior art search (if you're doing/paying for one), filing fees, and other government fees. Government fees are typically a minor component of the cost; attorney fees are usually the biggest chunk of the outlay. But it's important to balance the cost of paying a competent attorney to draft and submit the application against what you might lose if your application is not done correctly—or at all.

Attorney Fees

Attorney fees for a patent application can vary widely, and there are different arrangements you can make for billing and payment, depending on the attorney you use. (For more information about fee arrangements, see Chapter 10.)

As a general rule, you can expect attorney fees for:

1. Prior art search, analysis of the search, and consultation regarding options
2. Preparing a provisional, utility, or design patent application
3. Preparing a response to an Office Action or an appeal
4. Paying issuance and maintenance fees
5. Reviewing correspondence from the Patent Office, reporting status to you, and consultation regarding options at each juncture in your case

Of course, these fees will vary tremendously from attorney to attorney. Also, consider that some attorneys prefer to not charge for the "in-betweens," such as reviewing an Office Action and discussing options with you. Instead, they make these costs part of the bigger fees, like preparing your patent application, responding to an Office Action, and so on.

USPTO Fees

The Patent Office bases its fee structure on the size of the entity requesting patent protection. The three-tiered system identifies *large entities*, *small entities*, and *micro entities*, based on the criteria below.

Large Entities

The only patent applicants that pay large entity fees are major corporations and institutions and inventors who have assigned or licensed their invention to a large entity (or have an agreement or obligation to license or assign their application to a large entity).

Small Entities

Good news for individuals: the government charges you less! USPTO filing fees are significantly reduced for small entities. In fact, most fees for small entities are about half those paid by large entities. This can save you hundreds and sometimes thousands of dollars. To qualify as a small entity, an organization must have less than 500 employees. Obviously, individuals and small businesses qualify as small entities. In simple terms, this means you might be eligible for small entity fees in the beginning of the process, but if, later, you make a deal with a large entity to pay you royalties for using your invention, you will instead have to start paying large entity fees.

Micro Entities

Micro entity status gives you an even better deal and is especially set up for first-time inventors. If you qualify, most micro entity fees are only a quarter of large entity fees! To be eligible, you need to have filed fewer than four (4) utility patent applications (ever), and your annual income must be less than a certain threshold. This threshold goes up each year. For example, in 2016, it was $160,971 per year. Visit patent-book.com/USPTO-fees for the latest threshold figure and information about the current USPTO fees.

Filing Fees

When you file a typical utility application, you pay a *basic filing fee*, a *search fee*, and an *examination fee*. In 2016 these combined fees, for a small entity, totaled $730. Filing a design application is less expensive. In 2016 it cost about $380 for a small entity to file a design application. At $130, the small entity fee for filing a provisional application is even less.

Issuance Fees

Once the Patent Office approves your patent application, and before they issue it—you guessed it—you gotta pay another fee! The small entity *issuance fee* for utility applications was $480 in 2016, and the small entity issuance fee for design application was $280.

Maintenance Fees

Once your utility patent is granted, *maintenance fees* must be paid to keep your patent enforceable against others. The fees are due at 3½, 7½, and 11½ years after the patent issuance date. The fee increases significantly at each interval, and small entity or micro entity status will save you considerably. The fees range from several hundred to several thousand dollars, and, if need be, they can be paid up to

Figure 11.1 Maintenance fees must be paid to keep a patent in force

six months late *with a surcharge*. If, however, the fees are not paid by four, eight, and twelve years after issuance, the patent will expire prematurely.

The consequence of an expired patent is that you will not be able to stop infringers. Many people use the upcoming maintenance fees as a time to reflect on the value of keeping their patent in force. Consider that *if* you are exploiting your patent in any significant way, the fees will seem *relatively* nominal. But if several years have gone by and you have not made progress with manufacturing or licensing, the fees will seem significant, and it might make sense to let it go and put your attention on something else.

Design patents do not require the payment of maintenance fees.

Extension Fees

Certain due dates for responding to the Patent Office can be extended by paying *extension fees*. For example, when the Patent Office issues an Office Action, they will ordinarily require a response within three months. This deadline can be extended up to three months by paying an extension fee. Note that, while a one-month extension fee is relatively cheap, a three-month extension fee is not. Also, the Patent Office cannot extend the deadline by more than three months.

You might notice in the fee schedule that there are fees listed for four-month extensions and five-month extensions, but these cannot be used to extend a typical three-month Office Action. In general, just because there is a fee listed doesn't mean it's possible to use it in your circumstance! Also, note that, when your patent application is approved, and you receive a Notice of Allowance, the time period for paying the issuance fee and supplying anything else that is required by the examiner in the Notice of Allowance cannot be extended.

Foreign Filing Fees

Filing foreign patent applications requires significant fees that will vary tremendously; these fees can be estimated by your patent attorney. Consider, however, that filing a Patent Cooperation Treaty (PCT) application will cost at least a few thousand dollars in filing fees alone. (For more on foreign applications, including PCTs, see Chapter 9.)

Tips for Saving Money

- *Do the initial research before hiring an attorney*. While having a professional patent search done might be wise before committing to have an attorney prepare and file a patent application, it pays to do some research on your own, too, before you hire an attorney. Sometimes, after a few hours—or even a few minutes—of searching, you will find your exact idea and decide not to proceed with a patent. Clearly, doing this before you hire an attorney will save you money.
- *Learn about the process*. (Check! You are reading this book.)
- *Make your invention descriptions clear and concise*. When the attorney does not fully understand your invention, it will probably require more

time for your attorney to clarify it with you. Even worse, the attorney might do significant work on your patent application, and only after you review the application do you both discover that the invention was misunderstood. All of that wasted effort can be very costly! Take the time to prepare a clear description and sketches. On the other hand, whenever a client has come to me with a clear, well-organized description, in my mind I can begin to visualize an easier path toward completing the patent application. It makes it seem more palatable, and typically results in a lower fee quote. I can think of plenty of times when clients have brought me a very clear and understandable description of a relatively complicated invention, which would have typically cost more. But in the end, because the description was so clear, I charged only an average fee rather than charging more due to the complexity of the project. They might not have realized it at the time, but they saved thousands just by taking a few hours to write something up!

- *Provide screenshots for apps in a usable format, and edit out branding and dates.* (But don't conceal prior use dates from your attorney!) If you are seeking to patent an app, good quality screenshots will not only save time and help the attorney better understand your invention but can also save you money on drawing expenses as well. Many attorneys try to avoid including things like logos or other brand names in drawings filed in a patent application. Ask your attorney for his or her preference and what's best in your situation. Most likely, if you can edit out logos and brand names to make the drawings more generic, this might make them more usable for the attorney. Also, many attorneys want to avoid screenshots that depict a particular date. For example, a travel reservation app might show an example of a user selecting flight dates. Perhaps the dates on your screenshots are old, as they were created while you were tinkering with the idea last year. Your attorney might want to carefully consider how to depict these to avoid creating a future issue by giving the false impression that your app had been in public use prior to filing. In any case, you still must be candid with the attorney about any prior public use, so that he or she can discuss any public use issues with you.

- *Have reasonable expectations.* Come across as someone who will have reasonable expectations and will be relatively low maintenance. In theory, your attorney should charge you an appropriate amount for the work he or she does for you. Consider, however, that if you come across as a client who will require more than usual from them—namely, more time, more attention, and more special or unusual requests—they will probably quote you higher fees, whether consciously or unconsciously. If they notice that you are the type of person who wants to do things your way, even in arenas where they are the expert, they will expect that it's going to take more time and attention with you than with other clients. However, if, instead, you come across as someone who will follow the program as outlined by the attorney and will be easy to work with, they will look at your project as an ordinary one that they can include in their workflow without much hoopla. I'll tell you that, in some cases, when I've done a project for a repeat client who was really easy to deal with the first time, I've actually lowered my fees a bit!

What You Can Do on Your Own

Clients ask all the time whether they can write a patent application on their own and then pay me to read it, dot the I's, and cross the T's. The few times I agreed to do this, at first the application they wrote looked well written, and I thought that I would just need to give it a good read and fix a few words and phrases. But then, as I got deeper into it, I began to notice more significant issues that would need to be addressed. Perhaps I first noticed that their claims needed to be rewritten. Then, I saw some other adjustments that needed to be made in the application to keep it consistent with the way I corrected the claims. Then I realized that the way the invention was described in the specification—the way it was structured and broken down into elements—was not compatible with the changes I needed to make to it. Now—already a few hours in—it was clear that I wouldn't be able to just edit their application in a few hours as we had discussed. To do it right I would need to rewrite it from scratch. At that point, I knew that what I needed to do would be way beyond their budget because it would be way beyond the "just spend a few hours reading and fixing it up" that we spoke about.

Over the years I've seen attempts made by very intelligent PhD scientists, by medical doctors, and by other *really* smart people, at writing a patent application. In the end, what they wrote was very useful for me to understand their invention—but would never fly as an actual patent application. I mean, I've trained quite a few new lawyers to write patent applications, and even after they've already drafted a dozen or so, I could easily spend as much time proofreading and correcting their work as it would take for me to write one from scratch! In sum, writing your application with the expectation it's going to be good enough to file, or even close, isn't a realistic option.

But if you do want to save money and streamline the process by putting in some of your own effort, consider what I suggested previously: research it on your own first, and be sure to write a clear description. The application that you attempt to write on your own won't be usable as an actual patent application for filing, but it will give your attorney a really great understanding of your invention when writing a patent application for you.

The same goes for creating good drawings. If you have the capability of producing 3D CAD drawings, great! The attorney will still need to have a set of patent drawings made, but your CAD drawings will help ensure that your attorney has a clear understanding of what you have in mind, and they will help streamline the process of creating patent drawings that are correct the first time.

Summary

Generally, the biggest component of the cost of patenting is the attorney fees. Regarding the government fees also required, small businesses and individuals get a discount. For a listing of the current Patent Office fees, visit www.patent-book.com/USPTO-fees.

Being prepared with a good description is the best way to save money on the process and help you make the most out of the services you are paying for.

Getting Your Idea Across

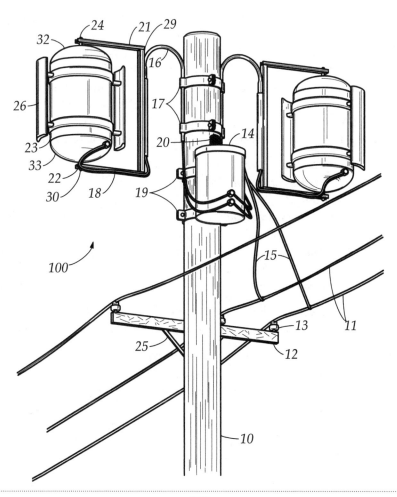

Communicating your idea effectively to the USPTO is the job of your patent attorney. However, explaining it to everyone else—including your patent attorney, in the beginning—is your job.

Why are we even talking about this in a book about obtaining patents? Because too many inventors have difficulty explaining their invention and can easily be to their detriment. When asked to summarize their invention, they might take twenty or thirty minutes to get to the point about what they've come up with. They might talk about the origins of the problem. They might talk about future ways they could enhance their idea. They might describe some of the pieces in great detail. What they typically fail to do is provide a short, concise overview. For instance, if they had invented the automobile, they might talk about the tires, about the limitations of horse-drawn carriages, about how gasoline combusts in the cylinder, about how to crank it to get it started, and so on. What they fail to say is, "My invention is a horseless vehicle that is propelled by a gasoline burning engine."

Communicating What Makes Your Idea Great

Whether you are meeting with a patent attorney for the first time, presenting your idea to a potential investor or possible manufacturer, or even trying to write your own patent application, it's all about *communication*. What matters most is getting the point across about what makes your idea different, special, and valuable.

It's a really good idea to practice describing your invention in simple, understandable terms. If you're going to be successful, you're going to have that conversation many times, with many people. In fact, it's fair to say that your success is a function of how good you get at communicating your idea to people simply, in a way that's clear and that piques the other person's interest.

- *The less clear you are when you explain your idea to your patent attorney, the less likely the patent application will embrace what's important about your idea, and the application will likely cost more.*
- *The less clear you are when you explain your idea to others, the easier it will be to lose their interest.*
- *The less clear you are when you explain your idea to investors or potential partners, the less confidence they will have in your ability as a business leader to understand what's important, focus, and get the job done.*

As an inventor who's going to seek a patent—and might also be seeking investors and venture partners—here are three important marketing tools you must develop:

- *An elevator speech—a short, no-nonsense description that you can deliver in twenty to thirty seconds and that makes it abundantly clear what your invention is all about and leaves your listener instantly curious about working with you.*
- *A written description—a longer but still no-nonsense description that completely nails what the problem is, how you solved it, and how your solution is different from—and superior to—the ways others have tried to solve the problem.*

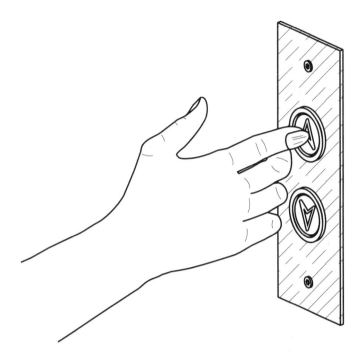

- *A more formal approach for presentations and marketing—your "elevator speech plus." It's like a one-minute infomercial that covers the background, how you came up with your idea, testing, success, and why you're sharing it.*

When you are working on this pitch, practice telling your story as you would to a close friend or family member—someone you would genuinely like to help out. Ask yourself:

What points would get their attention?

What would really help them understand the superiority of my solution?

What could I say about my own experience that would convince them to try my idea?

Write down all the answers you come up with to those questions, and then narrow down the list to just the most universal, attractive, and engaging answers. (Keep a copy of the full list, though—you'll need it later.)

Now you're ready to build your story.

The Elevator Speech

Once again, your elevator pitch is a short, no-nonsense description that you can deliver in twenty to thirty seconds, which makes it abundantly clear what your invention is all about and leaves your listener instantly curious about working with you. It should quickly answer the question of who it is for, what problem it solves, and what is the solution that your idea provides.

Of course, not everyone who hears your pitch will become an investor or a partner. Your patent attorney will not be investing in your idea; the way you tell

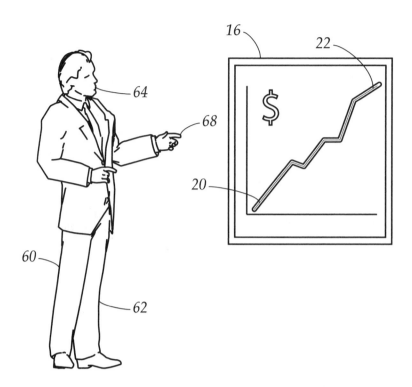

your story, though, will be a great help to him or her in knowing what you are looking for, what questions to ask, and where to start. As for the rest of your audience, generally speaking, anyone who's serious about the niche in which your idea fits—and particularly people who already do business in that niche—will be interested in a well-told story about a new idea that works. And they are likely to know other people who will also be interested.

The Written Description

Using your elevator speech as an outline, fill in the details with the other ideas you came up with earlier. Expand on them. Paint an even clearer picture of your idea, how it works, and why it's better. This is where you can tell of the choices you made in determining the precise problem, what's lacking in the prior art, the importance of any details (materials, manufacturing methods) that make your solution superior, and the excitement of the people for whom it worked!

Depending on the nature of your idea, you also might use drawings, flowcharts, schematics, wireframes, animations, a model, or photos of a working prototype to communicate the nature of your invention. Your written description will be very helpful—not only to your patent attorney, but also to any artists, designers, or manufacturers you enlist to create these elements for you.

The More Formal Approach (for Marketing and Presentations)

Your elevator speech is intended as a tool—to help you make contacts and set up meetings where you can give a fuller, more detailed presentation, whether those meetings are for one or two potential partners or for a roomful of potential investors. When you do meet with them, they will want to hear a lot more than your twenty-second overview. Here is where your full list of ideas comes in.

You probably won't need all this when talking to your attorney, and it's not really necessary for making a description, but it is helpful to have it readily available, just in case. This presentation contains the elements of the written description, plus the enthusiasm of the elevator speech. It's likely that you would expand this presentation with PowerPoint, video, detailed cost and market analysis figures, and/or a prototype of your invention.

Here's the basic outline of a format that I find effective when seeking to influence others about the value of your idea:

1. I was just like you! I had this problem.
2. I tried everything.
3. One day, I discovered the secret.
4. It worked!
5. I tried it out with friends. They had great results too!
6. Now I want to share it with you!

If you look at this format, you'll begin to notice just how many product pitches, marketing materials, and "sales letters" you've heard, seen, and read that follow this format. Why? Because it works. It allows other people to relate to you, relate to your story, trust that you have the solution, and consider that this is their opportunity to get involved.

Once again, going through the exercise of creating these tools is partially about having them available to actually use and partially about enhancing your ability to effectively communicate your idea. Your audience will vary, so the more skilled you become at getting your idea across, the better you will be able to shape your message to fit the audience when necessary.

More Advanced Tools for Getting Your Idea Across

Have you ever noticed that when you explain something to others, you end up understanding it better yourself? While this is a good reason to write a description, it's also a good reason to invest in some other tools that will help you to both better understand your invention and make it easier to convey to others.

Building a Prototype

A *prototype* is not necessary to get a patent. When people think about getting a patent, the old folklore has them think about a guy going down to the Patent Office with his contraption. He sits in the waiting room with other inventors, all with their contraptions on their laps. Each time it's someone's turn, he or she goes into a room with a long table and several judges (kind of like that scene from *Flashdance*), and demonstrates his/her invention to the judges. Before reading this book, was that your perception as well? The truth is, working models have not been required by the Patent Office since 1880. So there's no room full of judges looking at your

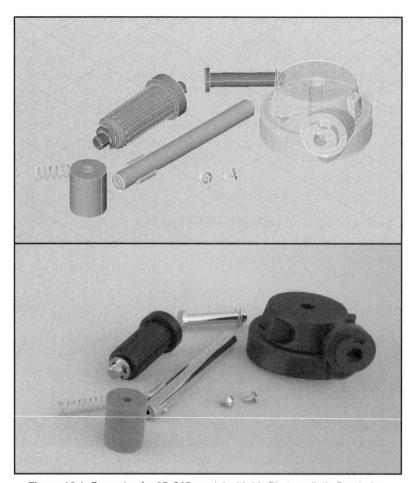

Figure 12.1 Example of a 3D CAD model with it's Photorealistic Rendering

invention prototype at the Patent Office. Everything is done in writing and through drawings that show what the invention *would be like* if it were actually built.

If a prototype is not necessary, then why build one? Building a prototype will help you to test out your notions about your idea. It might seem like a good idea, and sound like it could work. Once you build a prototype, however, you get to see whether it really works as expected, and you might get ideas for modifications that would make it better.

Also, a physical prototype is probably the most effective tool you could create for getting your idea across to someone you want to invest in your product, license it, or otherwise believe in what you are doing. It's one thing to describe your idea to someone; it's quite another to actually give it to her so that she can hold it in her hands and see for herself. A physical prototype is probably the best sales tool you can hope for when pitching your product to others.

While building a prototype is a great step to take, sometimes it can be prohibitively expensive and impractical. In those cases, there are other ways to prototype an idea without actually building the physical invention. When it's practical and economical, however, it often pays to build a prototype.

3D Models, Photo-Realistic Renderings, Virtual Prototypes

Creating drawings, videos, and animations that show others what your invention would be like is probably the next best thing to having a physical prototype. A good 3D artist can bring your idea to life and show its possibilities in a way that would impress even you, the inventor. These are great tools for getting your idea across and for helping you further design your product. Accordingly, we'll talk about these methods in more detail in Chapter 13—Developing Your Idea.

Talking with Investors

When most people talk to an investor about their idea, they believe it's their job to sell the investor on what a great opportunity it is. In most cases, this leads to behavior that will quickly turn the investor off—not only to the idea but to you as well.

Consider that potential investors can quickly smell an opportunity without you overselling it. The next thing they are sniffing for is whether they can trust you to pursue it and trust you with their money!

I can't tell you how many conversations I've heard where the inventor starts out with, "I want to tell you about my billion-dollar idea," or even, "my invention is worth $200 million."

Understandably, the inventor thinks this will get the investors excited. In reality, however, what the investors are thinking is some combination of:

- *Yeah, right!*
- *Does this guy have any idea what it really takes to make a product that successful?*
- *How do I get out of this conversation politely?*

This is where overselling it will work strongly against you!

It's good to know numbers for conversations with investors. But the types of numbers you should know are:

- *The investment capital that will be needed, both upfront and as the company grows*
- *The cost per unit for producing the product*
- *The size of the market*

Note that the size of the market could often be a number like $200 million or one billion. But talking about the size of the market is very different from calling it a billion-dollar idea! When you do, it's like you are declaring that you can step in and claim the whole market, *just like that!* No investors want to give their money to someone who believes it will be that easy, or that certain. It's good to be confident, but it's better to have humility about the outcome—while having confidence in your ability and determination to solve problems as they arise along the way.

In sum, the way to impress investors is to demonstrate that their money is safe with you. They will know it is when they see that you are capable, knowledgeable, and skillful, and yet you have enough humility to realize you don't know everything. They want someone who can look to the sky with his or her feet firmly on the ground.

Summary

Communicating with others about your idea and your vision for it is perhaps the most important skill for any inventor/entrepreneur to cultivate. It's the key to inspiring investors, venture partners, and even consumers. Your ability to get your idea across, and instill confidence in others that you can manifest your vision, is how you create a future that includes your idea and your business. For more information, visit the companion site for this book: www.patent-book.com.

CHAPTER

13

Developing Your Idea

In Chapter 1, we talked about why you want a patent, and how it pays to take a closer look at your own reasons before jumping in. In this chapter, we are going to talk about what exactly you want to patent, manufacture, and market, and how it pays to take a closer look at that before jumping in too far.

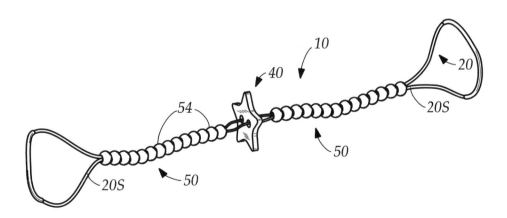

Often, the idea for a new product strikes like lightning. You are doing something, perhaps something you do every day, and you realize: *If I just had this thing, it would be so much better!* And suddenly, you came up with an idea for this thing that would make it so much better! Immediately you wondered—does something like this exist? If it doesn't exist, then why didn't anyone think of it? So you decide: *This is it! Let's patent it, make it, and sell it!*

Other times you're using something someone else designed, something that's already on the market, and you're not happy with it. Something about the way it works (or doesn't work), or the way it feels in your hand, or the awkward way you have to use it makes you think: *Why didn't they make this differently? How come it goes like this instead of like that? Why doesn't it just work easier?* So you decide right then to design *something better.*

If you fall within the first category, your invention came quickly, and you might think, "This is it—this is my invention." But before jumping right into patenting and manufacturing it, consider slowing the process down a bit. Consider trying the second route and spending some time *developing* your idea.

The Four Factors for Success

The decision to develop your idea usually leads to a slower, more deliberate process. Surely people would appreciate a product that works better and is easier to

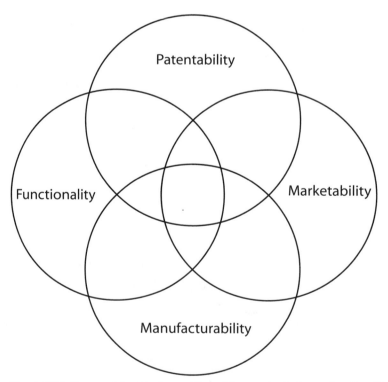

Figure 13.1 Successful products are often at the overlap between functionality, marketability, manufacturability, and patentability

use! Well, generally speaking, a lot of time, effort, and planning go into the design and development of a product.

There are four key factors that contribute to a product's success: *functionality*, *marketability*, *manufacturability*, and *patentability*.

Functionality

Functionality considers how well the product solves the need or problem. Optimizing functionality for a particular solution means considering whether there is a more effective solution, and whether any modifications would make it more effective.

Marketability

Marketability considers how effectively the product matches the *perceived* needs and wants of consumers. What is the solution that consumers are *already seeking*? Marketability considers how many potential consumers exist for your product. It also considers competing products and solutions from others, how much consumers are willing to pay for the others, and how much they would be willing to pay for yours.

Manufacturability

Manufacturability considers what it would take to actually make your product. Is it possible to manufacture it? What materials would be necessary and what would it cost?

Patentability

Patentability considers what aspects of your product can be protected. Are these the same features that make the product functional and marketable?

These are *factors*, in that they work together to contribute to a product's success— that is, a product can sometimes be weak in one category, yet still be successful if the other factors are strong. For example, if a product is strongly marketable and functional but weakly patentable, it can be very successful, but you will likely have to share the market with copycats. If a product is strongly marketable, it can perhaps be less functional (not quite the best solution), yet people will still want it.

Of course, all of this is a simplification and optimizing any of these categories is way more complex. The key here is to be sure to look at the different areas that affect a product's success. I've seen people neglect one of these areas, to their detriment. Probably the least considered category is manufacturability. Lots of patents exist for products that just cannot be manufactured, or the cost to manufacture them would be way more than the market would ever pay for the finished product.

Steps to Developing a New Product

Identify the Need

The first step when developing a new product is to *identify an important need*. Typically, ideas spring from situations where the need is obvious to you at the time. However, it's important to consider the need from every angle you can think of, so you're not too limited in your approach throughout the design process. And one of those angles is to consider not just how you experience this need, but how other people experience the need. What solutions do other people use or try? How satisfied do you think they are with the currently available solutions?

Brainstorm Solutions

The next step is to *brainstorm new solutions*. The first idea or two may have come to you at that "lightning moment." But it's important not to stop there. Brainstorming means coming up with lots of possible solutions and listing them—without rating them, judging them, or ruling them out, yet. So keep looking for other solutions. When you give the need your full attention, you'll probably see that there are other ways to solve it besides the one that crossed you mind first. Make notes, sketches, whatever helps you to capture every possible way you can think of to satisfy the need. You might even consider how the existing products do it. Perhaps the best solution is not a dramatically new product, but just a clever modification of one that already exists.

Explore Each Solution

Once you have your list, take a closer look at each possible solution. Now is the time to start rating them. You can decide on your criteria for rating the ideas on your list. The criteria can be things such as: which solution would seem to be the most effective, the quickest and easiest to implement, the simplest, the least expensive, the most appealing to others, and so on. Using these criteria, you can pick which one or which ones seem to provide the best opportunity to pursue further.

Make a Drawing

Everyone who has kept a notebook and made drawings of their ideas knows that drawing is more than just a way to get it on paper—it's part of the design process. As you are drawing, details come out that you might not have considered before. You should always spend time making sketches.

Most people will resist making a drawing, insisting that they can't draw. Much of the time when people resist making a drawing for their idea, however, it's because, on some level, they know that there's a part of it they haven't figured out yet. So, if you haven't figured out how the pieces will come together, you'll resist making a drawing of it. Consider, however, that making a drawing is a great opportunity to figure out how the pieces connect.

If you spend just ten minutes on a drawing, you'll likely express the things already in your mind. If you spend an hour instead, I guarantee that there will be something new on that paper that you hadn't thought of before!

Prototype and Test Your Possible Solutions

Once you have a few ideas for how you'd solve the problem or need, pick the one you *think* would work best. Then, *prototype and test* your solution: build it, code it, carve it out of wood—or whatever is appropriate to your invention—and try it out. Does it work? Could it work better?

Usually, the testing phase of the process births more ideas. You may decide that a different material (or coding language) or a slightly different approach might yield better results. This is where *iteration* comes in. You make changes to your original idea as you see what works and what doesn't.

My good friend Thom Wright is an extraordinarily skilled artist and product designer. (In fact, he created the drawings you see throughout this book.) Thom has created prototypes for many different product ideas, as well as the example at the end of this chapter. I asked him whether iteration is important for the development process. He told me simply, "There is no substitute for iteration. Professionals do it. Amateurs do it. Anyone who comes out with something that *actually works* has iterated their idea through at least a few versions."

There's no telling how many iterations you'll go through before you hit the version that you are satisfied with. The results or a test might even point you toward a different solution—possibly one that you'd thought of in your brainstorming session, but that you had previously ruled out. The folklore is that Thomas Edison tried 10,000 different ways to make a light bulb before he hit on a workable and commercially viable solution. Whether he actually tried that many or not, the point is, he tried lots of different ways before he settled on a solution. Remember his often quoted statement, "Invention is 1 percent inspiration and 99 percent perspiration," and keep going until you find the approach that works for you.

Investigate the Marketability, Manufacturability, and Patentability of Your Preferred Solution

Once you know you have a functional invention, it's important to do market research to find out what people are looking for. Find out what other solutions to the problem are already out there. Search online, go to stores, and talk with other people about how they solve the problem and whether the available solutions meet their needs. *It's typically possible to do all this without discussing your ideas with others!* You'll want to find out more about what's currently available, what customers *wish* they had available to them, and how much they would be willing to pay for an effective solution.

It's also important to determine the manufacturability. After figuring out whether it would be feasible to manufacture your product, you want to get a rough estimate of what it would cost to make it.

This is how you can begin to find out whether there might be a market for what you've come up with. Would enough people be willing to purchase your product, considering what it would cost for you to make and sell it? If you realize that there's no market for it, you might move on to your next idea. But if there is a market for it, the research you do now can be helpful in refining your product for the marketplace, and possibly for selling the idea to investors or venture partners. And of course, knowing whether there is a market will be helpful in deciding whether it's worth pursuing a patent.

When you have a workable solution, and it looks like there's a market for it, it's time to find out whether your idea is patentable and, if so, whether a patent is worth the cost to you. Now that you have spent time considering possible solutions to satisfy the need, you can make a more informed choice about whether a patent would be worthwhile.

I won't get into the details of determining patentability here because that information is covered earlier in this book. The thing to look at is whether your best solution is different enough to be patentable. Also, consider how much better your solution is when compared with other available solutions. Remember that, for preventing competition, patenting your solution might only be worthwhile if it makes your product more compelling in the marketplace than other available solutions. How good is your solution in the categories we talked about? Is there an overlap among functionality, patentability, marketability, and manufacturability?

Case Study

As an attorney with a duty of confidentiality to my clients, I don't usually get to share their stories with the public, let alone provide insight into their development process.

Thom Wright, the patent draftsman and designer mentioned earlier in this chapter, is a dedicated prototyper (you can see some of his work at www.inventionart.net). Since most of his client work is also typically confidential, he generously agreed to share the story of how he developed one of his personal projects. Thom's story provides a good example of how your initial concept of your invention, and any assumptions you may have made about it, can easily shift when you prototype it.

As an avid stand-up paddle boarder near his home in Western Massachusetts, Thom often found himself boarding in breezy conditions. Thinking about wind surfers, which are sails mated to small watercraft, he wondered about the possibility of somehow having a sail available when he is paddle boarding.

His first idea was to combine the sail with the paddle. He thought: *the sail would be useful, but it should also not get in the way of paddling when the sail was not being used.* So how about having the sail deploy from the shaft of the paddle? In this way, the paddle could be used like any other paddle. Then when you want the sail, you press a release button, and within seconds, you have a sail to harness the wind. He began making drawings and building a prototype of this concept.

Thom also had me investigate the patentability. I ordered a patent search, and he and I went over the results together. As it turned out, he wasn't the first to think of the idea of attaching a sail to a paddle. But the idea of having it deploy from the paddle when needed, and fold into the paddle shaft when not in use, seemed quite different from what others had proposed in the prior art. We filed a provisional patent application to secure his priority while he continued to develop it.

As he built his prototypes, he encountered various challenges. Some parts could be 3D printed, yet others were complex mechanisms, and he sought to find off-the-shelf parts that he could use. Eventually, he was successful in building

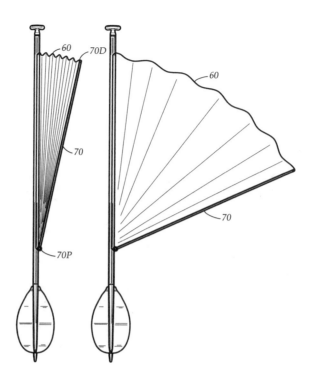

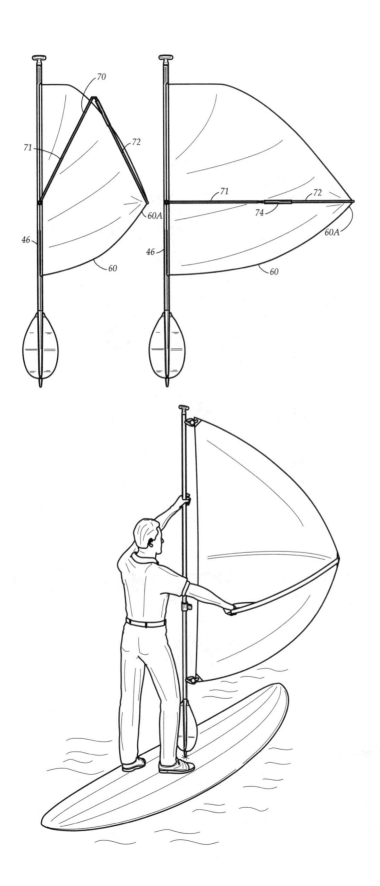

a working prototype. One of the things he realized when testing his prototype, however, was that he wanted the sail to be bigger. But the sail could only be so big and still fold into the shaft.

At a certain point, Thom realized that he had *made an assumption* that the sail should deploy from *inside* the paddle so that it would not be in the way of paddling when the sail was not being used. But what if it didn't need to? What if the sail could stay outside the paddle and still not be in the way?

He began exploring solutions that didn't require folding the sail into the shaft. After several prototypes, he created a modified version, which had a sail that remained outside the paddle shaft, but folded neatly into a bag when not in use. Because it did not need to fit within the paddle, the sail could be much bigger. It did not require a complicated and expensive mechanism to deploy, and it was much simpler to manufacture at an appropriate price point.

Now that Thom had a solution for the product he wanted, we took a look at the patentability. As it turns out, from the prior art search we did previously, attaching the sail to the outside was probably not different enough to be patentable, but was quite functional, marketable, and manufacturable. Since we had only filed a provisional, not much was lost when we decided to abandon it.

He currently manufactures and sells this product on a relatively small scale that is just about perfect for a side project. The important lesson here is that he didn't get trapped into following his first idea—even when all indications were that he should pursue a different direction with the product. And more importantly, the fact that he filed a patent application on one version didn't compel him to follow through on manufacturing that version when it no longer made sense to do so.

Summary

For any given invention that produces a particular result, there are typically other ways to produce the same or similar result. It pays to take the time to look at alternative solutions, alternative structures, and features before committing to pursue the invention as you originally conceived it. Whether you believe you can draw or not, you should always spend time drawing your invention, since most often the act of drawing it will give you new perspectives.

For more information about developing your invention, visit www.patent-book.com/invention-development.

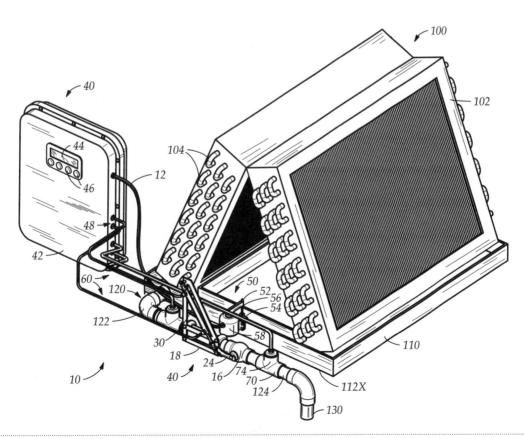

CHAPTER

14

Having a Plan to Profit from Your Patent

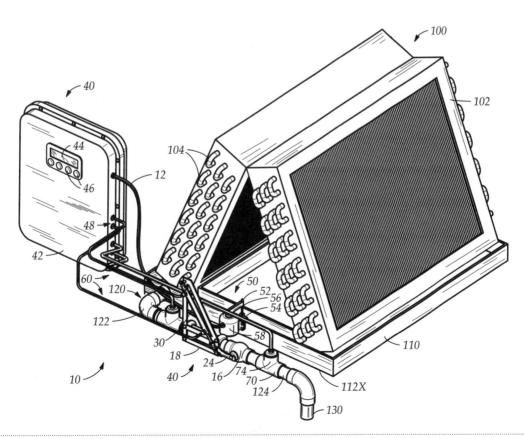

"If you build a better mousetrap, the world will beat a path to your door."
-Attributed to Ralph Waldo Emerson

I'm sure you have heard that phrase before. It can be very inspiring to people endeavoring to build the "better mousetrap" in their field. The problem is, just building the better mousetrap is not enough!

The biggest issue that inventors face is not getting a patent; it's finding a way to *monetize* their patent—that is, how to make money with it, how to profit from it, how to recoup their investment. I believe the biggest reason for this is that they think the above quote is true! They believe that once they have the idea, and protect it, that the money will naturally follow. As a result, they fail to have a plan to monetize their invention.

Most often, not having a plan won't necessarily lead you to failure, but it won't lead you anywhere good, either. I often say there is only one reason inventors fail: it's because they stop. Until you stop, until you give up, you haven't failed. But without a plan, without a direction, it's easy to get stuck. And once you get stuck, the only thing that will get you unstuck is something that indicates what to do next. There is nothing more effective at shining the light on your next opportunity than to have a plan—or at least some sense of the direction in which you are headed.

The main lesson is that your patent will not monetize itself. No one will beat a path to your door and pay you for your patent or even tell you what to do next. If you want to make money with your patent, you need to make a plan—and you don't need to reinvent the wheel to do so. You can borrow from the approaches that others have used to successfully monetize their inventions.

Let's consider, then, different ways that you can plan to monetize and profit from your patent. Each monetization strategy has its own *business model*. Let's take a look at what makes a business model so that you can begin to select one that fits your project.

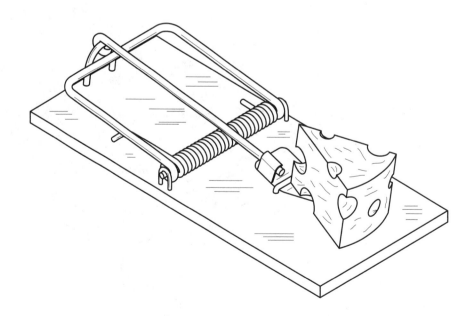

Exploring Different Business Models for Your Idea

Your business model is the blueprint for your business. If your business were a machine, what would be the working parts and how would they connect? Your business model is the plan that illustrates who your customers are, what it is that you provide to them, how they pay you, how you deliver it to them, and what it costs for you to deliver it. Or, said differently, it is a plan for how you make money from a given product or service, or both.

For example, the business model for a retail store is: you purchase products at wholesale prices, and then you sell them at retail prices to customers who walk into your store. The revenue is derived from product sales. The main costs for delivering these product sales include the wholesale cost of the products, the cost of maintaining the store (including rent, utilities, etc.), the labor cost for people to work in the store, and advertising (so the customers know what you're selling). Your profit is the difference between what you sell the products for and what you paid for them, minus the costs of running the store.

As another example, to see how even a simple business model has a lot of room for creativity, consider a business built around product sales generated on a website. In its basic form, revenue is again derived from product sales, and the costs for delivery include the cost of building and maintaining the website, wholesale cost of the products, and shipping costs. If you chose this business model over the retail store model, you are banking on the fact that the cost of building the website, getting people to visit it, and shipping products to them will be less than the cost of having a physical store.

But then, consider that even if you choose the online store business model, there's significant flexibility in that business model. Among other things, you might vary how you actually fulfill the product orders. You might, for example, maintain a warehouse. In that case, your costs will comprise the cost of the warehouse facility (including rent, utilities, etc.), the cost of maintaining unsold inventory, the cost of warehouse labor, and so on. You might also consider a model in which you don't maintain a warehouse; instead, you have the products "drop-shipped" directly from the manufacturer or an order fulfillment company to your customers, once they place their orders on your website. In the latter case, you would probably pay a fixed amount to the order fulfillment company every time a product is shipped. When you are selling a small volume, this fixed cost would definitely be cheaper than maintaining a warehouse. If you sell a lot of products, however, warehousing and shipping the products yourself may save you money in the long run.

Considering this, you can begin to see how much room for creativity there is in configuring the business model for your enterprise. These areas include:

- who your customers are,
- how you get your customers,
- what value (product/service) you deliver to your customers,
- how you deliver that value,
- how the customer pays,
- what business resources you need,

- what the main activities your business needs to perform "in house" are and which activities you will outsource to others,
- with whom your business needs to partner for the outsourced activities, and
- the costs of doing so.

This looks like a lot to consider, but how you fill in the blanks determines the model for how your business will operate—and how you'll make money from your patent.

Comparing the Two Most Common Business Models Used by Successful Inventors

Most often, when an inventor thinks of pursuing a new product idea, two very basic business models come to mind: *manufacturing* and *licensing*.

Manufacturing

Manufacturing is a very simple business model that any kid with a lemonade stand understands. Make the product, sell it for more than it cost to make it, and you end up with a profit. Manufacturing a new invention requires that you get involved in every aspect of developing, manufacturing, and selling the product. In essence, you're starting a business. If you are not already in business, in the field of your invention, the major question to ask is, *do you want to be?*

Licensing

Licensing as a business model is also quite simple. You find someone who's willing to manufacture, market, distribute, and sell your product. In turn, that person or company pays you a royalty for the right to do so. The advantage of licensing is huge: someone else does everything needed for developing, manufacturing, and selling the product. The main challenge with licensing is finding someone who both appreciates the value of your invention *and* is in a position to develop, manufacture, and sell it.

Levels of Risk

Whether you choose to license or manufacture your product will clearly result in different amounts of responsibility for getting the product out there. But it also will result in a different level of risk—and a different profit potential. When you are licensing your product, you take on very little risk, but when you manufacture it, you assume lots of risk in terms of time, effort, and capital.

Consider a product that might sell for $10 retail. The wholesale selling price would likely be about $5, the overall manufacturing cost perhaps $1.50, and a reasonable royalty on such a product would be perhaps $0.50. As a result:

- If you licensed the product, you would make the reasonable royalty of $0.50.
- If you manufactured the product, you would make the difference between the wholesale selling price of $5 and the manufacturing cost of $1.50. Thus, you would make $3.50.

Compared to licensing, then, if you manufactured the product, you would make *seven times more!* But, you would also need to take on a lot more responsibility and assume *all* of the risk for the success or failure of the product.

Other Monetization Approaches

There are several other ways to make money from a patent without manufacturing or selling a product.

Selling Outright

A patent can be sold, and the entirety of rights conveyed to another. The downside of selling a patent is that the price paid to you will usually be a rather conservative (low) figure, based on the fact that the purchaser knows he or she is assuming all of the risks for whether the patent will prove valuable. On the other hand, if the patent proves to be very valuable—if the product sells beyond all expectations—the inventor already got paid and won't get to share in any of that success!

But consider also that the opportunity to sell your patent outright is rarely an option, except in cases where the value is very clear to the purchaser. Otherwise, rather than paying you a sum of money up front to purchase the patent, he or she would probably rather license the patent from you, paying you a royalty for units actually sold, and then see how it goes.

Litigation

Some companies build a business model based on litigation—that is, the entity either develops or acquires patents with the sole intention of bringing patent infringement lawsuits against others. Companies like this are sometimes called "non-practicing entities," or NPEs, because the company doesn't actually make products based on its patents; it uses them offensively. In recent years, the tide has turned somewhat against NPEs, making it more difficult to win big judgments, and thereby making this business model less viable.

Aggregation

Realizing that patents tend to be more valuable when part of a group of patents, some companies go into business to create patent portfolios. The company buys patents and *aggregates them* into a more valuable group. Often, aggregators focus on a rather specific niche, such as telecommunications. Sometimes the aggregator will license the group of patents to multiple companies within the field, to serve as a defense against NPEs that might buy them and sue. Other times, the aggregator may sell the group of patents for a profit.

As a Defensive Strategy

When a patent is related to your business, sometimes the monetization strategy is to keep a portfolio of patents that serves as a defensive tool when others within your industry assert their patents against your company. Even if you have no intention of suing others, when they happen to come knocking with their own lawsuit, the patents you own that they happen to infringe can be an effective bargaining chip for you.

As Loan Collateral

In recent years, various companies have sprung up that offer creative financing packages to businesses and use the businesses' patents as collateral. Sometimes this money is intended specifically to finance litigation; other times, it's just to help a company finance operations and expansion. While this type of financing is probably less available for start-ups or for patents without proven value, it might be part of your monetization strategy down the road.

As an Exit Strategy

Again, you might not intend to make a product based on the patent, or to license it. When it's time to sell the company, however, and you own a portfolio of patents that are closely related to your industry, it can greatly increase the value of the company.

Building a Patent Portfolio

Some of the approaches described above become easier, more viable, and more profitable when you own several patents, as opposed to just one. A *patent portfolio* can be a great thing to have for several reasons:

- You can realize an income stream from licensing fees on more than one patent. It gives you a greater ability to "hedge"; as the revenue from one patent dies off, the revenue from another picks up.
- Should you decide to sell your company, a revenue-generating patent portfolio is an asset that adds to the company's value.
- Having several patents tends to be much more valuable for licensing, and when litigating, than just having a single patent.

You can create a patent portfolio by filing patent applications on different ideas and innovations within the same field. Your patent portfolio can also stem from a single patent application, where later innovations were added as a continuation-in-part, and where multiple inventions from the original patent application lead to divisionals, as we discussed in Chapter 9. For most companies, the patent portfolio will be a combination of the two: more than one patent "family," with several patents that arose from each family.

More about Licensing

The concept of patent licensing is simple: someone pays you to use your patent. The most typical arrangement involves the payment of royalties for each unit sold. In a standard licensing agreement, the arrangement is "non-exclusive." This means that if you make a licensing deal with company A to pay you royalties for products it sells, you can also make an agreement with company B to pay you royalties for products *it* sells.

Sometimes a company will want to be the only one that can sell your product. In such case, that company would want to enter into an "exclusive" licensing agreement. Typically, the terms should be a little more favorable to you—as an incentive to make the agreement exclusive.

However, there are some traps to avoid in exclusive licensing agreements. The biggest risk is that the company "shelves" the product—i.e., drops the ball in marketing or decides to not produce your product at all. If payment in the agreement were based solely on units produced or sold, the company would have no obligation to pay you. Even worse, if you gave the company the exclusive, the agreement would prevent you from working with another company to pick up the ball—and probably would even prevent you from manufacturing it yourself! Generally, a good licensing agreement will provide a solution for this scenario, such as by requiring a minimum royalty guarantee and a means for you to escape the contract and go elsewhere if the minimum is not met. For these and other reasons, exclusive agreements must be entered into carefully. A skilled attorney will know how to minimize the risk and prevent you from getting stuck in a bad exclusive licensing arrangement.

Even with licensing as a business model, there is much room for variation. If you are intent on licensing a single patent to a single company, perhaps it's relatively simple. But if you believe your technology might be licensed to multiple companies within an industry, or might concern multiple patents across several industries, then you must design and build a licensing company to achieve that.

Business Model Design

Even within the two basic business models, licensing and manufacturing, there's still much more to figure out, and much opportunity for your creativity. Business model design is an activity that's worth spending time on. It gives you the opportunity to investigate different options for monetizing your invention and your patent.

Business model design is an examination of the numerous possible ways to run a business and to make money with any business concept. I highly recommend that you take the time to research effective business models as you consider inventing and patenting. It's a great way to explore your options and to make the right plan for getting your idea out into the world.

The Business Model Canvas

A tool you may find useful for designing, understanding, and revising your business model is the Business Model Canvas, developed by Alex Osterwalder and Yves Pigneur in Lucerne, Switzerland. This tool has been purportedly adopted by a million business model designers and business school students around the world. What makes it such an effective tool is that it takes the concepts of a business model and makes them visual.

Figure 14.1, below, shows a blank Business Model Canvas. The nine spaces on the canvas represent the key elements of a business model and provide room for brainstorming different possibilities. On a larger version, which you can download and print from the Strategyzer AG website, you can quickly put down ideas using self-stick notes, discuss them with others, and easily move them around to try out different possibilities.

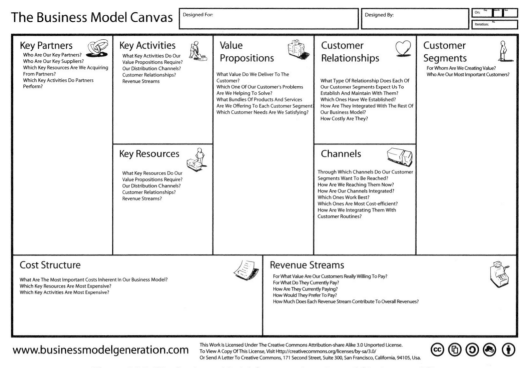

Figure 14.1 The Business Model Canvas – (courtesy of Strategyzer AG)

It's so important that everyone involved in a business understands how the business operates and what they, as individuals, are contributing to its success. Ordinarily, however, this is not the case. Typically, the concepts that make up their business's model are discussed back and forth by executives and their employees, or by entrepreneurs and their venture capitalists. Because these discussions mostly take place in a piecemeal fashion, without much structure, it's rare that any two people in an organization have the same understanding of the business model. When you make it visual, however, something magical happens: all of a sudden, everyone working together is on the same page about what makes the business model tick.

If you are committed to having a well-thought-out business model strategy that is fully understood by you and the people you are working with, I highly recommend that you check out the "Business Model Canvas" at the Strategyzer AG website: www.businessmodelgeneration.com, or read their book, *Business Model Generation.*

Summary

The only reason inventors ever fail is because they stop. Until they stop, they have not failed! Having a plan for monetizing your invention is the best way to have a direction, and thereby avoid stopping when you get a little stuck. Explore possible business models for pursuing your invention to help find a path and a vision that fits best for you.

Look to the companion site for additional information about business models and tips for monetizing your invention: www.patent-book.com/plan.

CHAPTER

15

Infringement: What to Know and What to Do

Like everything else in this book, the point of this chapter is to help you understand the principles and dynamics of patent litigation—not to substitute for the advice of an experienced attorney. I emphasize this now because, when it comes to litigation, there can be a lot more at stake—and therefore a lot more at risk—than when you are just seeking a patent.

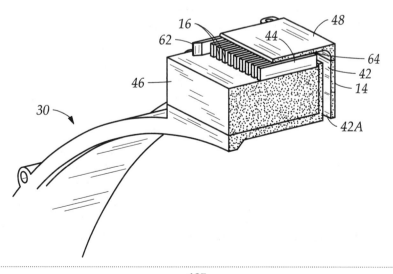

NOTE

In this discussion, "you" refers to the person being accused of patent infringement. Please don't take offense at this! It's just a useful device for explaining how patent infringement works (and when it doesn't work). One reason is that understanding how someone might escape your web puts you in a better position to strengthen it. Also, if you move forward with your product—or with practically any business that experiences success—there's a fair chance someone might allege that you're infringing his or her patent. You may as well get used to it.

Generally, the biggest risk of a misstep during the patenting process is that you may not get the patent. This means you might waste your money or not earn money that you possibly would have if you did get the patent. So when you're seeking a patent, it's hard to lose more money than you actually have—or than you make a conscious decision to spend—because, when you've taken it as far as you would like to go, giving up is always an option. And the costs typically stop when you say "when."

With litigation, however, it *can* cost you more money than you have—and there isn't always a "stop" button when you want to get off the ride! So, if you find yourself facing patent litigation, be ready to get professional advice. But in the same spirit as the rest of this book, I want you to make good decisions and use your counsel wisely. Here are some principles to help you understand the dynamics of patent infringement, and when it may be a real problem for your business.

When Might You Be an Infringer?

You can only infringe a patent that has not yet expired. For utility patents, this is typically twenty years after the first utility application was filed, as long as the maintenance fees have been paid. When a patent is part of a family of patents, the patent will expire twenty years after the first utility patent application was filed in the United States from which priority is claimed. In making this calculation, don't count any provisional or foreign applications from which priority was claimed. For design applications, the patent expires fifteen years from the patent issuance date if the patent application was *filed* after May 2015, and otherwise expire fourteen years from the patent issuance date.

What May Infringe?

To infringe a utility patent, you need to violate the claims of the patent. (For a full understanding of patent claims, see Chapter 6, "What Goes into a Patent Application?") With an *apparatus claim*, this means that your product contains all of the claim limitations in at least one of the independent claims in the patent. With a *method claim*, it means that your product, process, or app meets all the method steps in at least one of the independent claims in the patent.

For a utility patent, you would *literally infringe* the patent if your product contains all of the claim elements of at least one claim in that patent. If your product doesn't exactly contain every element, you still need to use caution, however. If you come close to *literal infringement* of at least one claim, and there is just an insubstantial difference between your product and that claim, you might still be infringing by the *doctrine of equivalents*.

For design patents, the test of infringement is: Does your product have a substantially similar appearance to the patented design in the eye of an ordinary observer. In design patent litigation, this would typically be up to a jury to decide. Note that, although finding infringement might seem straightforward when two products look quite similar, it's limited by two things: functionality and the prior art. For example:

> Two shovels might look alike because they each have a handgrip, a handle, and a spade. But shovels have those features to perform their intended function. What's more, many prior art shovels have these features. So, when determining whether a shovel is substantially similar in appearance to a patented shovel design, the jury would be instructed to look beyond those features that are functional and are in the prior art.

If this is confusing to you, there's good reason, and it would probably be confusing to a jury also! One of the benefits that a design patent holder enjoys is that it's hard to predict how a jury would find on this "substantially similar appearance" test. Because it's hard to predict what would happen if a potentially infringing design were ever tested in court, holders of design patents know that others will think twice before copying their design.

How Might You Be Infringing a Patent?

When it comes to utility patents, you need to actually make, use, or sell the infringing product. Infringement can also occur from offering to sell or importing an infringing product—even when no actual sale took place. As a practical matter, however, we typically talk of "making, using, and selling" as the prohibited acts.

With design patents, however, simply displaying the design can get you in trouble. An unauthorized display that shows the patented design can be a problem, even if you didn't manufacture, use, sell, or even offer to sell a single unit.

Another thing to know is that you cannot be an infringer *simply by patenting* an improvement on a patented product. As we discussed before, if someone else has a patent on a core technology, your improvement on that technology may be worthy of a patent. And as we said: even though you can patent it, you still wouldn't be able to manufacture it without the permission of the patent holder for the core technology on which it is based. On the other hand, if they see your patent, they can't sue you simply because you have invented a product that would infringe if manufactured. You would need to actually make, use, or sell the infringing product in order to become an infringer.

Who Can Be an Infringer?

Considering that infringement is about making, using, or selling something, when it comes to products, an infringer can be almost anyone in the manufacturing and distribution chain. It can be the manufacturer, the distributor, a wholesaler, a retailer, a salesperson, and even the consumer. As a practical matter, patent owners typically go after those with deep pockets. But that's not always the case!

The patent owners know it would be harmful to the manufacturer's sales relationships if the manufacturer's customers (wholesalers/distributors/retailers) were sued. The patent owners also know that if this happens, the manufacturer's customers will look to the manufacturer to "work it out" as quickly as possible. Accordingly, in many cases, a patent owner will go after wholesalers, distributors, and retailers as leverage against the manufacturer—to put pressure on the infringing manufacturer to settle the suit.

It becomes more complicated when infringement would require more than one party to perform all the steps in a patented method. This is especially common with software patents, patents on apps, and so forth. For example, a patented method might have one step that is performed by a user on his or her smartphone, together with other steps that are performed by a server in the cloud that is maintained by another entity.

When, exactly, patent infringement can result from steps performed by more than one actor has been the subject of much debate—and sometimes, conflicting court decisions. Before you decide that you're in the clear because you're only doing one portion of the patented method, or you're only selling part of a patented product, you'll want to consult a litigator about the possibility of *contributory patent infringement* and *infringement by inducement*.

As we discussed earlier, it is possible to infringe a patent, even if you obtain your own patent on an improvement. Here's an example:

> Barbara invents a chair having four legs and a seat. Imagine that this combination (four legs and a seat) is new and not obvious at the time Barbara invents it. Barbara is able to patent her chair.
>
> Soon, Charlie invents a chair with four legs, a seat, and a backrest. Charlie files a patent application. The Patent Office considers the backrest to be a non-obvious improvement on Barbara's chair. The Patent Office grants a patent to Charlie for the chair with four legs, a seat, and a backrest.
>
> How can this be? Doesn't the Patent Office care that the four-legged chair was Barbara's invention?

The answer is that the Patent Office *doesn't consider infringement* when determining patentability. That is, as long as Charlie's chair meets the requirements for patentability, the USPTO will grant the patent. Questions of infringement are not considered by the Patent Office; they are determined by the courts—and *only* if Charlie manufactures and sells his version of the chair *and* Barbara chooses to sue.

The bottom line, then, is that Charlie can get a patent, even though he would still be infringing Barbara's patent if he actually produced his version of the chair. Most people don't realize this scenario is possible. That's why they think getting a patent means they have the right to make their product!

In this case, if Charlie wants to make his chair, he must get permission from Barbara. But if Barbara thinks Charlie has the right idea with the backrest and wants to start manufacturing her chair with a backrest, she can't do that, either! She would need Charlie's permission because he holds the patent on the backrest idea.

Why Might You Infringe?

In some areas of IP—copyright, for example—the "why" matters. If you independently wrote a song that's very similar to a copyrighted song you had no access to and never heard before, it's not copyright infringement.

With patents, however, the "why" is mostly irrelevant. You may have toiled alone in your garage to invent a product, only to discover that someone else already has a patent that covers your product. The fact that you came up with it on your own is no defense. If your product is covered by their patent, it doesn't matter that you didn't know about their patent.

If you *do* know about the existence of another patent, and you continue to infringe it anyway, that can be serious. Unless you have a good-faith belief that you are not actually infringing their patent, your knowledge of the patent might be found by a court to be *willful*. If a court finds that a patent was *willfully infringed*, you can be ordered to pay additional damages.

Where Might You Infringe?

To infringe a US Patent, infringing activities must occur within the United States and its territories. But remember, any of the three activities—*making*, *using*, or *selling*—within the United States can constitute infringement. And add *displaying* to that list if it's a design patent. Understand, then, that manufacturing a product overseas won't get you out of trouble if you intend to sell the product in the United States.

When it comes to infringing products that are imported into the United States, the International Trade Commission (ITC) also has jurisdiction. Some patent owners choose to bring an action before the ITC, seeking an order that instructs the US Customs Service to halt infringing products from entering the country at its borders and ports. But you can't get an award of damages at the ITC, just an *injunction*.

Defenses to Patent Infringement

In almost every patent infringement lawsuit, the sole question isn't just, is your product covered by their patent? Generally, the defendant (the person being accused of infringement) will raise and argue one or more defenses. If the defendant is successful with certain of these defenses, then even if it's proven that his product infringes the patent, he can win the case!

Invalidity

The number one defense against patent infringement is to show that the patent is *not valid*. Quite simply, this means that the Patent Office should never have granted the patent in the first place. If the patent is found to be invalid, then naturally you won't have to worry about infringing it. Note that in a court of law, patents are *presumed* to be valid. If a patent has been granted by the US Patent Office, courts will assume that the Patent Office was correct in granting the patent, without sufficient evidence to the contrary. So the burden will be on you to show that the patent is invalid.

There are two main reasons for *patent invalidity*:

1. Prior art. Finding prior art that shows the invention existed previous to the patent is the most often used method for invalidating a patent. The rules are quite complicated about what types of previous inventions, publications, patents, and so forth, are prior art to an inventor and should have been considered by the examiner when deciding whether to grant a patent. Collecting any examples you know of that show potential prior art will be very helpful to your attorney when advising you.

2. Public disclosure. If a patent application was not filed within one year of the invention's public use or availability for sale within the United States, then the inventor should have been barred from getting a patent. If you have proof that, for example, the product was on sale more than one year before the patent priority date, this can be used to invalidate the patent (and escape liability).

As a practical matter, asserting that the patent is invalid can occur directly as a defense in the lawsuit, and also by filing a petition at the Patent Office requesting that a *reexamination* take place, or by initiating some other post-grant review procedure.

This is where talking strategy with an experienced patent litigator is important. Like a game of chess, often, the strategy created in the beginning, and each move that follows, will play a huge role in determining the outcome. Now, I've spent my career obtaining patents for clients, and I'm no expert in litigation strategy! So I asked Scott Stimpson, a colleague of mine who is a *patent litigator* at Sills Cummis & Gross P.C. in New York City, to explain the best way to defend an infringement action when you think the patent being asserted against you is invalid. This is what he told me:

> In any patent litigation, there are a number of considerations in determining how best to challenge a patent's validity. For instance, in court there is a presumption of validity and you must establish invalidity by "clear and convincing evidence." In Patent Office proceedings, however, there can be no presumption of validity or clear and convincing burden, making it easier to get the Patent Office to invalidate a patent. Depending on the type of proceeding in the Patent Office, however, there can be some downside risks, such as limited involvement by the patent challenger, or broad estoppels preventing other invalidity challenges if the Patent Office proceeding fails.

So there you have it. This stuff can get really complicated, really quickly! If you find yourself in this situation, definitely consult an expert—not just any old patent attorney, but a seasoned patent litigator.

Expiration

While it's extremely unlikely anyone would actually try to sue you over a patent that has passed its natural patent term (because you then would assert its expiration as an obvious defense), it pays to understand that if a patent has *prematurely expired*, before the end of its term, then your infringement of it *is also irrelevant*. As we noted previously, US utility patents require that the inventor pay maintenance fees to keep the patent in force. If these fees were not paid by the four-, eight-, and twelve-year anniversaries of the patent issuance date, the patent would expire early and could not be enforced against infringers.

You can check the maintenance fee status of a US patent on the Patent Office website. You'll need to know the patent number and the serial number; the serial number can be found on the front page of the patent document.

In some circumstances, the fee can be paid late, and an expired patent can be *revived* if the inventor pays a petition fee. Generally speaking, the petition (and the late payment) must be made within two years of the fee's original due date. So, if you see that a patent is recently expired for non-payment of maintenance fees, you must still use caution.

It's also important to check whether there's another patent by the same inventor. The inventor may have obtained more than one patent related to the product. Perhaps the one you are looking at was less important, and the inventor allowed it to lapse while still paying the maintenance fees on another, similar patent— meaning that there might still be an in-force patent that you need to watch out for!

No maintenance fees are required for design patents; no fees need to be paid to keep them in force for their full term.

Inequitable Conduct

As we discussed earlier, a person applying for a patent has an obligation to be truthful with the Patent Office during the patent process. This includes the obligation to tell the Patent Office about prior art and other facts that might influence the Patent Office's decision in granting the patent. If you can demonstrate, however, that the patent holder intended to deceive the Patent Office during the patent process, this can result in the patent claims being unenforceable and can effectively serve as a defense to patent infringement.

Equitable Estoppel

If you had a reasonable belief that the patent owner would not enforce the patent, perhaps based on promises from the patent owner, this may provide a defense to patent infringement. Of course, the facts and circumstances matter greatly in determining whether the patent owner's actions would provide a valid defense in your case.

Litigation—the General Process

While you might expect to receive a *cease and desist letter* before being sued, it's quite common that the patent holder will simply file a *patent infringement lawsuit*. In either event, the patent holder is claiming that infringement has already occurred— that someone has copied protected IP without the patent holder's consent.

Cease and Desist Letters—Why They Are Rarely Used

Just a decade or two ago, cease and desist letters were a more common way to notify an infringer of potential patent litigation and to try to get the infringer to stop before litigation became necessary. This method is now considerably less common—for good reason.

Typically, when one company felt it was being infringed by another, the company would have its lawyer draft and deliver a cease and desist letter. The point of the letter was to say, "Hey, we've got a problem here. If you don't stop, we'll sue." The intent was to get the other party to back down and either stop infringing or enter into a license agreement.

Frequently, these letters were used by people who had no intention of suing. Oh, they were willing to make the threat, but if the other side called their bluff, they were not ready to spend the money to bring a lawsuit in federal court!

The problem with cease and desist letters—and the reason this method fell out of favor—came when the people who received the letters turned around and sued the sender! How could they do this? Well, it's called a *declaratory judgment action* (or "DJ action"). By starting a DJ action, the letter recipient can sue to have the court declare that he or she is *not* in fact infringing—and, possibly, that the patent is invalid. Letter recipients can do this because the courts recognize that recipients have a right to settle the question of whether infringement is actually occurring rather than have the suspicion of infringement forever "hanging over their heads."

For a little guy who's just sending out letters, hoping that the allegedly infringing companies will cooperate, this can be devastating! Once the other side sues for a declaratory judgment, especially if he or she is seeking to invalidate your patent, you can't just drop it. You now need to spend a lot of money to litigate that question. If you dropped it—and didn't respond to the lawsuit— chances are that the other side would get a *default judgment* invalidating your patent. Yes, skilled patent defense lawyers do have some tricks up their sleeve for attempting to end an unwanted declaratory judgment action. For example, if a patent owner finds herself on the receiving end of a declaratory judgment action, she might issue a "covenant not to sue." Her promise not to sue the other party for infringement thereby settles the question of infringement that the other party claims would be "hanging over their heads" and may bring an end to the suit. But this is certainly not a situation you want to end up in!

Without knowing this could happen, some people will still send a cease and desist letter that they wrote on their own (without a lawyer) to a manufacturer. When a manufacturer sees this, the company might smell an easy win and file a DJ action. The person who sent the letter obviously had no idea this could happen—and would likely end up begging for mercy.

Nowadays, cease and desist letters are really only used by those who are ready to go to war over the infringement, who are very careful with the language to make sure it is not at all threatening, or who doubt the recipient would put up the money to file a declaratory judgment action.

The Mechanics of a Patent Infringement Suit

How patent infringement suits are handled is typically the subject of not just ordinary books but multi-volume sets of books! To be brief, and just in case you find yourself in this situation, let's take a 10,000-foot view of a typical patent infringement lawsuit.

The Start of the Suit—the Complaint

Without getting into too much detail, the complaint sets the facts as the patent holder sees them, providing the patent holder's version of the story. The complaint will specify whom the patent holder is suing ("you," in this example), which patent or patents the patent holder says are being infringed, and some facts about how you are infringing.

The complaint will typically come as a "summons and complaint," filed in a federal district court, and it will require that you file an answer to that district court within a certain time period or risk default. If you are ever served with a complaint, without a doubt, it's time to get a patent litigator involved.

The Answer

Within a specified time, the answer must be filed. This is your opportunity to deny the allegations by the patent holder. It's also a chance for you to assert various defenses. And if warranted, it's also an opportunity for you to assert counterclaims, which may include any "beef" that you have against the patent holder! If, for example, the patent holder happens to be someone you did business with and he or she didn't pay you or broke a contract with you, you could assert a counterclaim against the patent holder for breach of contract.

Motions to Dismiss

At various points during the litigation, you might bring a motion to have the court rule on some aspect of the case. This includes a possible *motion to dismiss*. Often early in the case, such as when the answer is filed, you might file a motion to dismiss—especially when the patent holder's complaint is deficient in some way. If you win, fantastic! The case is dismissed. If you lose the motion to dismiss, then the case continues.

Discovery

This is the opportunity for both sides to request facts and evidence from the other side. These requests can be for documents, to answer specific questions (called interrogatories), or to question a witness (called a deposition). For people who have never been involved in a lawsuit, being asked to answer all types of questions and to provide business documents to their adversary can be rather surprising and unsettling. They typically ask their lawyer: "Can they really do this?" And for me personally, it is one of the main reasons I don't do litigation; it can get really ugly and nasty during discovery.

Summary Judgment

A fundamental principle of our judge and jury system is that the duty of resolving questions about how the law is applied rests with the judge, and the duty of resolving questions about the facts rests with the jury. The main reason for a trial is typically to get all of the facts out, so that the questions can be resolved.

Sometimes, when there is no genuine dispute about the facts, and the only question is how the law should be applied to the case, one or both sides might request "summary judgment." If the judge agrees that there is no genuine issue of fact, he or she might rule in favor of either side, which could potentially end the suit without a trial.

Trial

When issues of fact still must be resolved, a trial is held. Witnesses are called, and evidence is presented by both sides as they attempt to prove their case.

Judgment

Following the end of the trial, a judgment will be rendered in favor of one of the parties—or sometimes, partially in favor of each party. Generally in a patent suit, the judgment will include an award of damages. The judgment may also include equitable relief, such as an injunction against future infringing activities.

Appeals

For many patent infringement lawsuits, the judgment isn't the end of the story! Often the judgment at the district court level is appealed. A peculiar thing about patent suits: no matter where in the country they take place, all appeals go to the Court of Appeals for the Federal Circuit (CAFC) in Washington, DC. If the parties are still not satisfied after a decision by the CAFC, they can ask the US Supreme Court to consider reviewing the decision.

Summary

Patent infringement is the flip side to obtaining a patent. When you are launching a product, it is conceivable that you might step on someone else's patent—or be accused of doing so. It pays to be able to tell the difference, because not every claim that you are infringing a patent will have a legitimate basis. Always get a professional involved if you get sued or when you are making steps toward launching your product and you think it might conflict with the rights of others.

For tips when dealing with a patent infringement situation, visit www. patent-book/infringement.

Alternative Protections— Trademark, Copyright, Trade Secret

Since the main focus of this book is obtaining a patent, so far we have concentrated just on patents. As a person pursuing an idea and possibly starting a business, however, there are other forms of intellectual property (IP) protection that might be useful to you. Once again, my intention is to give you the main concepts.

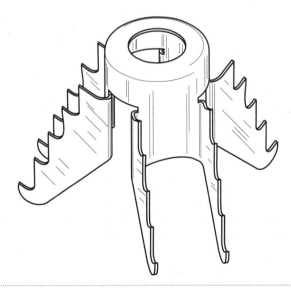

In this chapter, I will discuss what you need to know about trademark, copyright, and trade secret law as you start a business around your invention. If you want to learn more about these areas, and even get into some nitty gritty details about patent law, I recommend you check out another book recently published by the ABA: *Fundamentals of Intellectual Property Law*, by Stephen McJohn and Lorie Graham. It's written for non-lawyers and is a more in-depth study of the various areas of intellectual property concepts than I provide here.

At this point, you understand that utility patents and design patents are the appropriate ways to protect the functional aspects of a product and the ornamental aspects of its design. Once you've investigated whether your idea is patentable, and you have taken steps to establish patent protection if appropriate, you'll next want to look at trademarks—especially if you are preparing to launch your product.

Trademarks

Trademarks protect aspects of your product or service that consumers use to identify it in the marketplace. The essence of trademark law is preventing competitors from doing things that would confuse the public into thinking their product or service is yours. Several different types of distinctive elements of your product or service can be considered a "mark" or trademark. For example, a trademark can be a name, a logo, a slogan, and sometimes even the shape of the packaging, the color scheme of the labeling, or the shape of the product.

Research Trademarks Before You Get Too Far

Imagine that you're about to launch your product. You've come up with a clever name and a logo for it, and you're about to build a website that describes the product and displays the product name and logo. Clearly, you'll want to get that name and logo protected so that other people can't just jump on the bandwagon and use your name and logo to sell their own product.

Even if you're not so worried about copycats, if you're about to invest money in launching a product under a certain name, you want to be sure you can use that name without infringing others. It's very possible that you can get pretty far into the launch process—believing you'd come up with a unique name and logo, then spending significant money on the website, product labeling, and promotion—and then suddenly, out of the blue, someone is yelling at you, "Hey! This is *my* name!" And they're demanding that you stop using it. At that point, whatever money you've spent to promote that name—including all the advertising you've done and all the goodwill you've developed—will be wasted. Not only that, but they might even ask you to pay damages!

You should have trademark research done at an early stage, before you start using the name, to investigate whether it's protectable and whether you're going to be stepping on someone else's toes by using it. If you have a lot on the line, it pays to get the opinion of a trademark attorney about whether there's a likelihood of confusion with other marks. If the research shows that you're in the clear, consider getting your mark protected.

Protecting Your Trademark

Probably the most common types of trademarks are names, logos, and slogans. It's most likely that, as you launch a product, it will be one of these identifiers that you'll want to protect. Trademark rights for a mark are specifically associated with a certain class of goods or services. It's possible for someone else to use the same mark in an unrelated field, as long as there is no likelihood that people would be confused between them. For example, Apple Computer and Apple Bank seem to coexist without consumer confusion.

There are two ways for a mark to acquire trademark protection. By using the mark, it acquires *common law protection*, and by filing an application to register the mark, it may acquire the protection of a *federally registered trademark*.

Common law trademark protection develops automatically as your product becomes well-known among consumers. The more consumers know of your mark, the more recognizable it is, the greater your common law trademark protection. If you wanted to prevent competitors from using your product or service name based on your common law rights, you would need to show that consumers already attribute that name to your product or service and would find your competitors' use of that name confusing. Common law trademark protection, then, only exists in geographic areas where you can prove that the use of a similar mark by others is likely to create confusion. If your trademark has only been used locally, protection *will not* extend to places where they have never heard of you.

A registered trademark is acquired by going through the process of trademark registration with the USPTO. Trademark registration will protect your mark throughout the United States, even in places where they have never heard of you!

Trademark Registration Process

Actual use of your mark is essential for the registration process to be completed. As it turns out, this is where patent and trademark law are quite different: for a patent, using the invention before filing may cause you to lose your rights. With a trademark, however, selling products with your mark is essential to establish your rights! It's possible to apply, however, when you haven't yet begun to use the mark. Thus, there are two different paths to obtaining registration of your name/mark by the USPTO:

- **Option #1, an *"Actual Use"* application** (when you are already using the mark):
  1. Have your name/mark evaluated to determine whether it seems to be available for your goods and services.
  2. Use your brand name in *interstate commerce*.
  3. *File an application for registration with the USPTO.*
  4. *The mark is reviewed by a trademark examiner.*
  5. If approved, and not contested by others, you will receive a certificate of registration and you may use the ® symbol with your mark.
- **Option #2, an *"Intent-To-Use"* application** (when you have not used the mark yet, but genuinely intend to use it):

1. Have your name/mark evaluated to determine whether it seems to be available for your goods and services.
2. *File an application for registration.*
3. *The mark is reviewed by a trademark examiner.*
4. If approved, and not contested by others, you will receive a notice of allowance, requiring that—*within six months*—you either use the mark or seek an extension for another six months.
5. Use your brand name in interstate commerce and file a *statement of use.*
6. Receive a certificate of registration, and you may use the ® symbol with your brand name.

Each trademark is unique and so is every situation. Your attorney will help you to understand the nuances and develop an individual and appropriate trademark and intellectual property strategy for your business.

Trademark Evaluation

The purpose of the trademark evaluation is to collect information from you about your brand and about the goods and services that you offer under the brand, and then to determine if the brand name you have chosen is available. Not every brand name can be registered on the federal register; your attorney will advise you as to whether your mark qualifies. He or she may also suggest how to modify your mark so that it can qualify.

Trademark Application

While it is definitely more feasible to do a trademark application than a patent application without an attorney, it is still not advisable. The application process itself is straightforward, but there is some strategy and nuance involved in how you portray the mark and how you select the appropriate goods and service class(es) that you are applying for. It doesn't take a lot to get the knack of it and do a reasonable job if you have been through the process a few times. If you get it wrong, however, you will need to start over—losing your filing fees and whatever competitive advantage you might have gained by filing your application when you did.

Bottom line: if doing this right is important to your business, have it done by an attorney. It is not very expensive to enlist the help of a trademark attorney. Professional fees for trademark filing are a fraction of the fees involved with a patent application!

Copyright

One of the biggest reasons to talk about copyrights here is to see how it doesn't apply to protecting ideas. Often, when someone hears you have an invention, they will say: "That's a great idea, you should copyright it!" This is incorrect advice for two reasons:

- An invention is the wrong subject matter for a copyright. Copyright law protects items in specific categories, including books, movies, songs, plays, works of visual art, sculptures, and so on. These categories of artistic works are all suitable for copyright. An invention is not.

- Copyright does not protect ideas. Copyright protects *artistic expression*. If an artist has an idea for a painting of a barn in front of a stream and forest, copyright law will not protect that idea. Anyone can create a similar painting of a barn. But the artistic expression—the way that particular artist executed the idea, with her choice of colors, brush strokes, and composition—is protected against copying by copyright law.

So even for the correct subject matter—e.g., a painting—copyright law does not protect ideas.

Copyright Infringement

The way a copyright is infringed is when someone *actually copies* your work. He or she must actually have had access to (seen) your work and copied it. If, however, the person *coincidentally* made a work that was very similar to yours, it's not copyright infringement because he or she didn't actually copy it. That said, when works are very similar, courts may presume that there must have been access, if it seems that the similarities can only be explained through copying. This is especially the case when a work is widely disseminated and thus widely known.

Obtaining Copyright Protection

First, it is important to understand that copyright protection attaches the moment your work is "fixed in a tangible form." Your rights under copyright law come into existence the moment your poem is written, your song is recorded, your painting is painted.

And additional step you should take is to put a copyright notice on all of your works. Putting a copyright notice on your work puts others on notice that this is your work, and that copying is not permitted. You don't need to file any paperwork to do this. Just add it at the bottom of your work:

© 2016 Your Name

You can take the further step of registering your copyright. A registered copyright has certain advantages, such as: it allows you to sue infringers in federal court, it makes additional damage available, and it allows you to record the registration with US Customs to prevent the importation of infringing goods. Getting a copyright registration is rather simple and involves filling out a form online with the United States Copyright Office at www.copyright.gov.

Trade Secrets

Trade secret protection is about protecting the important secrets of a business—whatever they may be. Such secrets may include product details, formulas, customer lists, processes, and so on. The key to achieving trade secret protection is simple: *you keep it a secret*. You invest in safeguards to prevent your secret from becoming known. If your safeguards fail, and it is no longer secret, then your protection is gone, *unless* the secret was disclosed or uncovered through improper means. So if your formula was stolen because of industrial espionage or because of a disgruntled employee, trade secret law will give you legal rights to do something about it. In that case, the safeguards you employed to keep it a secret become evidence that you intended to keep it a secret, that it is something valuable, and that your company would be damaged by the misappropriation of this information.

Since trade secret protection relies on your ability to keep something a secret, it must be the type of thing that cannot be "reverse engineered." If an engineer could figure out the details of your product by taking it apart, or a chemist could determine the ingredients of your formula by analyzing it, then it's not suitable for trade secret protection. Generally, trade secrets must be something that is not widely known by the public, and not readily ascertainable.

In a sense, trade secret law is the opposite of patent law. Patent law requires that you fully disclose your idea to obtain protection. Trade secret law requires that you keep it secret. For this reason, it's typically impossible to have both patent protection and trade secret protection for the same thing. As a product develops, after a patent application has been filed, however, newly developed aspects and features need not be disclosed to the Patent Office and can be maintained as a trade secret (if they can be kept secret). With regard to the core invention, however, most often a choice must be made.

The most famous story about making a choice between patent and trade secret is the story of the Coca-Cola formula. When the Coca-Cola formula was invented by Dr. John S. Pemberton in 1886, he could have patented it. Patenting it would have required that he disclose the formula in his patent application. It also would have meant that the patent would have expired and the formula would have been in the public domain by around 1903. Instead, he decided to keep it as a trade secret. This was a wise move, because the company has managed to keep it secret—and thus, protected—for more than a hundred years past when the patent would have expired!

Whether trade secret protection is appropriate for your invention, or for aspects of your business, really depends on whether it's possible to keep it a secret. Since most products can be reverse engineered, and most new features you might add are probably easily recognized just by inspecting it, this is rarely an option for product ideas, or even for software. It's an important protection to know about, however, since, as you build a business, there will undoubtedly be know-how and other aspects of your business operation that you want to keep secret.

Summary

After you have protected your concept and other product features with patents, consider where copyright and trademark may apply. The artistic expression contained in certain aspects of your product, your website, and so on, may be protected with a copyright. A trademark is useful to prevent others from using your name, logo, and other distinguishing features that might confuse consumers. Trade secret is sometimes appropriate to protect aspects of your business that are important, if they can be kept secret.

Patenting Software and Apps

How Should Software Be Protected—with a Copyright or a Patent?

Even with the patentable subject matter controversy that we will discuss below, a patent is still the appropriate way to protect a process inherent in software and apps. What you would aim to protect are the processes that embody the distinctive functional features of your software product that make it unique in the marketplace.

As you know, when you're writing code, you could write it in many different ways. You could fulfill the same objective by writing code in different languages, and with lots of different techniques for organizing and grouping instructions. The final product—the code you wrote—could be literally protected with a copyright. But then again, knowing that anyone could vary significantly how the code is written, without changing the way it functions at all, this doesn't help you very much. At best, it protects against direct piracy of the code itself.

Copyright is often considered a secondary form of protection, as it's good for protecting things like the graphic design and images that are part of the user interface. A patent, however, is still the best way to protect the distinctive functionality and performance of the software.

At What Point During Development Should You Consider Patent Protection?

You should look into patent protection as early as you possibly can in the development cycle. The main reason you want to consider protection at an early stage is that a lot of what you find out about the prior art can shape what direction you go with development, and can rightfully influence whether further investment is appropriate.

Initial review of the "patent landscape" surrounding your app idea should be done before significant investment has taken place. Granted, you might already be beyond that. You might've already conducted significant work and moved toward launching the product—or actually launched it already. In a lot of cases, it's still not too late. It's really never too late to assess where you are and what you have. You might not necessarily file a patent application, but you should at least find out where you are by having the right type of evaluation conducted. And when a significant sum is being invested in development, that might even mean conducting infringement or Freedom to Operate evaluations.

Things happen quickly when developing apps. But remember that getting a good patent application prepared and filed can take some time. For various reasons, when possible, you should file your patent application(s) before you launch. And it's certainly safest to have an application filed before exposing your idea to people you don't personally know and trust, such as potential investors (or even other people on the development team).

Current Controversy over Patentable Subject Matter

In Chapter 2 we discussed how software and apps can often be protected as a process. This has been the case for the past two decades, where the series of steps inherent in operating and/or executing computer software is considered a process, and thereby provides patentable subject matter.

While this remains true, it's important to understand that the question of when patentable subject matter is present for software inventions has been the subject of significant debate in recent years. And since the US Supreme Court decided the case of *Alice Corp. v. CLS Bank Holdings* in the middle of 2014, the issue of patentable subject matter has gotten a bit more complicated and made the prospect of obtaining certain types of software patents less predictable.

After the *Alice* case, the USPTO and the courts began to reject many types of software patent applications as "abstract ideas" and therefore lacking patentable subject matter.

In the *Alice* case decision, the Court refused to define what exactly constitutes an "abstract idea" and instead referred to a series of past cases in which the Court had created a "judicial exception" [to patentability] when it ruled that patentable subject matter was lacking. The Court then gave Patent Office examiners the task of comparing the claims in the patent applications they are reviewing to these "judicial exceptions" and considering whether they have an abstract idea. Then, if the examiner thinks the claims include an abstract idea, the examiner is supposed to look and see whether the claims include "something more" that brings the claim beyond a mere abstract idea!

Do you find this confusing? Well, it seems that many examiners do as well! And many patent attorneys are reporting that the Patent Office examiners seem to be erring on the side of caution—issuing a 101 rejection (the invention lacking patentable subject matter) more often than not—and hoping that the patent attorney will respond with an explanation of how his or her client's invention *is not* an abstract idea!

Most commentators expect this situation to settle down in the coming months and years. Most also believe that examiners are interpreting *Alice* way too strictly and labeling software processes "abstract ideas" more frequently than the Supreme Court intended. With more clarity from the courts, examiners at the USPTO will be more likely to start taking favorable action in allowing more and rejecting fewer software patent applications. There are numerous cases pending in the courts, which will provide an opportunity for more clarity from the bench.

Even between the time I am writing this and when this book is printed, there is a fair chance this situation will change and—I think—improve. I promise to give you an updated status on the companion site for this book at www.patent-book.com/software.

As of this writing, however, the situation is more confused than it is settled. As a result, if you are considering patenting software, you should pay extra attention, and be especially strategic in determining *whether*, *when*, and *how* to file your patent application. For example:

- Your patent attorney can draft your application with *Alice* in mind— being sure to include detail and physical aspects to make the claimed invention less abstract, and providing "something more" that can help get beyond the examiner's concerns.
- The current climate may lower your chances of being approved for certain types of software patents. You should consider this carefully, and get the advice of a patent attorney versed in the current state of debate regarding patentable subject matter for software patents, when deciding *whether* to file a patent application.
- Because many software patents being examined today are likely to receive a 101 rejection stating that patentable subject matter is lacking, it might pay to try to put off examination as long as possible. Since many believe that the pendulum will eventually swing back in favor of patentable subject matter for software, there is a fair chance that you will be better off having your application examined later rather than sooner. At the very least, this means that speeding up examination through the "Track One" program would probably not be helpful to your chances. It also means that filing a provisional application—to delay examination by up to a year—may be a worthwhile strategy.

Finding the Right Patent Attorney for a Software Patent Project

You may think it's of primary importance to hire a patent attorney who "specializes" in the kind of technology your idea uses, but that's not necessarily a good idea. Here are a few reasons why.

Conflict of Interest Considerations

Generally, people who have experience writing patent applications in the exact field of your invention will not be able to handle your case because of a *conflict of interest*. Other times they may shy away from handling your case, even without a true conflict of interest, but because of a "business conflict"—meaning that they have a large client who would not be happy to know that the attorney was representing someone else in a similar technology field.

Depth of Focus

To me, writing a broad patent application is more about being able to *see the forest* than being able *to see the trees*. The "trees" are important to an extent, and must be described sufficiently to fulfill certain requirements under the rules of the USPTO. However, whether someone infringes your patent or not is all about whether the person who wrote your patent application had enough vision to also see "the forest."

This is where I get a little opinionated about what it takes to write effective patent applications! After all, we're in the appendices of this book. It's kind of like you hung out after the main speech and are listening to some of my more private musings.

In my experience "technical experts" are so accustomed to their technical field that they focus their patent applications on highly technical details and perhaps overemphasize and spend too much time on smaller technical differences. A skilled patent attorney without the benefit of all of those details, however, will tend to focus more on the big picture differences and create a patent application that describes the differences that are important to—and recognizable by—those people who matter: consumers, manufacturers, and courts of law.

I find that the patent attorneys who write the best applications are not necessarily the specialists in the technical fields in which they write. Instead, they possess a critical (and rare) skill for immediately cutting through the fluff and understanding what matters most about your idea.

This is *especially* true if you yourself aren't an expert in the field. What is most important to convey about your invention is the *concept*—the part that *you* conceived of, with your limited technical knowledge in the field. As long as your patent attorney has a greater ability than you do to explain your invention in technical terms, this "non-expert" patent attorney will likely emphasize the concepts that you consider to be your invention anyway, and not get stuck in some technical details that even *you* won't understand when you read the patent application the attorney writes. When it comes down to it, it's *you* who needs to read and understand the application before it's filed. After all, your name and signature will be on it!

Is a Deep Technology Specialist Worth the Premium Price?

Generally, the only place you find the deep technology specialists is at large firms. The cost of obtaining a patent at a large firm could be considerably greater than elsewhere. Even if you are willing to pay the price, however, many large firms won't even take on a smaller client. Which makes sense—when the typical firm client is spending a million dollars a year or more in legal fees, the firm is probably not set up to handle the smaller ones efficiently. Working with large firms comes at a premium price. Often, however, individuals and small companies don't have access to large firms. And when they do, they are seldom treated as well as the firm's big clients.

For a more in-depth discussion of how to choose a patent attorney, see Chapter 10, "Working with a Patent Attorney."

Summary

Patents remain the most effective way to protect software and apps. Patentability research should be done at an early point in the software development process, at which point initial protection might be established. Follow-up filings can be made as the software evolves, with care taken to provide patentable subject matter in the "post-*Alice*" software patenting landscape. Since the "patentable subject matter" controversy is constantly evolving, be sure to find out the latest at www.patent-book.com/software.

Marketing Your Idea

When people say they want help "marketing" their invention, what they often mean is that they want help finding a company willing to buy or license their patent and take over manufacturing, sales, and distribution of the product. If this is what you want, you should consider the discussions below about licensing and selling patent rights, which in some ways are similar to what we discussed in Chapter 14, "Having a Plan to Profit from Your Patent." If, however, you're looking to explore how your product fits in the marketplace, how it fits the needs of consumers, and how to effectively communicate that message, the next part of this discussion is for you.

Marketing

For the purpose of this discussion, let's say that "marketing" is about understanding what people need and ensuring that your product is a match for their needs. More accurately, it is about what they perceive or believe that they need and how well your product meets those perceptions.

Whether it's a major appliance or a mobile app, a lot of time, effort, and money go into bringing a new product to market. Fundamentally, all of that effort and investment should only be put into a product that the marketplace (individuals or businesses) both needs and wants. To know that you're headed in the right direction, with the right product, it's critically important to know your market. To start, you must have a clear picture of the *who*, *what*, *why*, *where*, and *how much*:

- *Who* needs your product?

- *What* do they need it to do?
- *Why* do they need it (as opposed to someone else's product)?
- *Where* will they find out about it and get it into their hands?
- *How much* will they be willing to pay for it?

The answers to these questions depend on your product, of course. But for every product, there are some commonalities that we can explore here.

Who

It's critical to spell out exactly who is the target market for your product. If it's a novelty item, the market may be eight- to twelve-year-olds who attend school, share interests with a large group of friends, and have pocket money to spend at their own discretion. If it's a new kind of bit for an industrial drill press, your market may be middle-management decision makers at small- to medium-sized manufacturing companies that have the budgetary resources for new equipment. Obviously, these two markets don't have the same needs, and they aren't likely to gather information—or see advertisements—in the same places. It may sound too simple, but identifying the correct target market for a product—*any product*—is a science, and it requires research.

So do your research. Look at other products that are similar to yours. Who buys them? Write up a short profile of your "ideal customer." You can tweak it later, as you collect more information, but knowing the general group you're aiming for is a good start in the right direction.

What

Now that you know whom you want to sell to, you need to know what they need your product for. What is the problem that your product solves for them? Is it . . .

- for entertainment?
- to make drying their hair easier?
- to freeze molded component parts to below-zero temperatures?
- to help them find a restaurant that will deliver a gluten-free pizza on a Sunday night?

A couple of sentences about what your product will do for your ideal customer, to make his or her work/life easier or more enjoyable, will help you define further the identity of your ideal customer—and may also help you refine the details of your product's design.

Why

This is a *big* question: Why would *anyone* be better off for having purchased *your* product, as opposed to any other solution currently available in the marketplace? Having done your prior art research, you should already be off to a running start on the answer to this question. Perhaps there's no other solution available. That would be the ideal situation for you, but it's unlikely. People are managing to get through life without your product. Perhaps the available solutions are too messy, time consuming, or expensive for your ideal customer. It's important to know at least a couple of *really good* reasons why he or she would be better off owning your product than any other out there.

Where

Location! Location! Location! Knowing where your ideal customer shops for products such as yours helps you in several ways, including offering insight into:

- which retailers or distributors may be likely to sell it,
- what delivery method will be required to get it into your customer's hands, and
- where you'll need to place your advertising to get your customer's attention.

It may be that you'll need to set up your own online store, buy banner ads on social media sites, and deliver your product as a download, or you'll need to purchase magazine ads, billboard space, or airtime on a popular television or radio program and arrange shipping to a retailer or directly to your customer's place of business. Or maybe you should create a podcast or an infomercial to post on YouTube. There are many, many possible answers to the "where" question. Understanding your ideal customer is the only way to narrow down the range of available options.

How Much

This question comes last because you need the answers to the other four in order to answer this one. You also need to know the manufacturing and packaging costs (if any) of each unit of your product. Determining how much to charge requires that you first crunch the following cost numbers:

- Raw materials
- Manufacturing
- Packaging
- Distribution
- Advertising

For a software product that's sold as a download, the first three can be ignored, but distribution costs would include server hosting and/or any percentage or fixed fee per download from wherever you sell it. And every product requires some kind of advertising, even if you intend only to "stir up interest by word of mouth"; managing any ad campaign—including a "free" one on social media—takes time. As they say, time is money.

Licensing or Selling Patent Rights

You may be seeking patent protection for your idea because you want to "sell" the rights to a big company and make money on the IP, without the hassles of manufacturing or distributing a product yourself. Or, your idea may be one that *requires* other patented parts to be used with it to make a viable product, and another company or individual already holds those required patents. Either way, if your idea is going to get into the marketplace, licensing or selling your patent rights may be the only way to achieve your goal.

As with marketing a product that you manufacture and distribute yourself, understanding *who* would be interested in the rights to your idea, *what* it will do for them, *why* they need your IP instead of some other IP that does a similar thing, and *how much* it may be worth to them are critical pieces of information for you to have if you're going to get the deal done. The answers to these questions will form the basis of your "pitch" to potential buyers. They generally won't just buy it; *you're going to have to sell it to them.*

Who

Researching the "who" part of a licensing deal or IP sale requires that you find out not only who would use the end product that contains your IP but also who would manufacture that end product. Which companies are making the thing? Which ones already have something similar, and which ones are looking to break into that market? An eager up-and-coming company may be more excited about your breakthrough; on the other hand, an established player might have deeper pockets and more interest in shutting out the competition.

What

Your target company needs your idea to . . .? The answers to this question and the next are the bases for your "elevator speech" when you "happen to bump into" an executive of the company you're targeting. You should be able to summarize—in thirty seconds or less—the usefulness and benefit of your idea to his or her company. The short paragraph you write up to answer this question will also form an important part of your pitch, when you do make the opportunity to present your idea to the decision makers at your target company. Understand further that if you are targeting more than one company, the answer will likely be different for each one. Be prepared. Do your research and *know* your target.

Why

The reason that your target buyer should buy from you instead of from someone else with a similar idea/patent must be crystal clear to the buyer—so you must be crystal clear on it first. There has to be a great reason that your solution to their problem will work better for them than anyone else's. *Find it.* The more solid your reasoning in answering this question, the more likely they are to say "yes."

How Much

The question of how much to ask for in return for the license or the patent requires *a lot* of research. You need to know whether your target buyer would prefer to license the IP or buy it outright. You also need to know—at least in ballpark figures—what percentage of the per-unit cost your idea would represent and how many units they're likely to manufacture. Or, if they would give away the end product (like a free app), how much they might be willing to budget to pay for it.

Often the entity that would be interested in purchasing your patent is the same entity that would potentially license it. If your patent is important to manufacturing the product they intend to manufacture, then they'll need to either purchase it or license it. Kind of like, if you want to live in a certain city, you know you need an apartment or a home there; then you can decide whether you want to rent or own. Similarly, the company that needs your patent might be interested in renting (licensing it) or owning (purchasing it).

All other things being equal, the potential purchase price when you sell your patent outright would be lower than the expected royalty earnings if you kept it and licensed it. This is because when you sell it, the purchaser takes all the risk of whether the product will succeed or fail. If the product has failed, the company already paid for the patent, as well as for all of the expenses involved in manufacturing and launching the product. For this reason, generally, unless the manufacturer is very optimistic about the success of the product, selling the patent to the manufacturer won't be an option.

When you license your patent to a manufacturer, you share the risk with that company regarding the success or failure of the product. As noted, this is usually preferable to the manufacturer. In addition, it gives you the ability to share in the upside. Since sales predictions will always be conservative, if the product is a hit, you'll have the opportunity to share in the success. If, however, you sold the patent, regardless of how successful the product is, you won't earn anything more because you already got paid.

Summary

Marketing your invention can mean trying to find a company that wants to license it or seeking to find a market for your product among consumers. Either direction requires the same basic inquiry: Is it a match? For whom? And what is the competition? The better you understand your marketing target, the more likely you will be able to make a deal or to sell your products.

APPENDIX

C

Utility Patent Example

US005997404A

United States Patent [19]

Sardo

[11]	**Patent Number:**	**5,997,404**
[45]	**Date of Patent:**	**Dec. 7, 1999**

[54] **RACKING SYSTEM FOR ARRANGING POOL BALLS**

[76] Inventor: **Louis Sardo**, 1629 Verdugo Blvd., La Canada, Calif. 91011

[21] Appl. No.: **09/221,724**

[22] Filed: **Dec. 29, 1998**

[51] Int. Cl.⁶ **A63D 15/00**
[52] U.S. Cl. **473/40**; 473/21; 473/26
[58] Field of Search 473/40, 21, 26, 473/73, 75

[56] **References Cited**

U.S. PATENT DOCUMENTS

464,475	12/1891	Sweet	473/40
1,052,461	2/1913	Chase	473/40
3,253,826	5/1966	Cook	473/40
3,992,005	11/1976	Richey	473/40
4,469,328	9/1984	Pacitti	473/40
5,376,054	12/1994	Kwasny et al.	473/40
5,601,495	2/1997	Silverman	473/40
5,735,750	4/1998	Silverman	473/40

Primary Examiner—Jeanette Chapman
Attorney, Agent, or Firm—Goldstein & Canino

[57] **ABSTRACT**

A racking device, for racking a plurality of pool balls upon a pool table have a felt top surface, comprising a rack for containing the balls, a pressure base located above the rack, and a sweeper frame capable of relative vertical movement with respect to the rack. A plurality of sweeper pins extend vertically downward from the sweeper frame, which each selectively engage one of the pool balls. A lever causes the sweeper pins to move downward toward the balls to engage the balls and urge them toward each other. A pressure pin is associated with each of the balls, which firmly press upon the apexes of the balls to press them downward into the felt top surface so that the balls hold their positions in a tight, racked formation even after the racking device has been lifted upward and away from the balls.

20 Claims, 4 Drawing Sheets

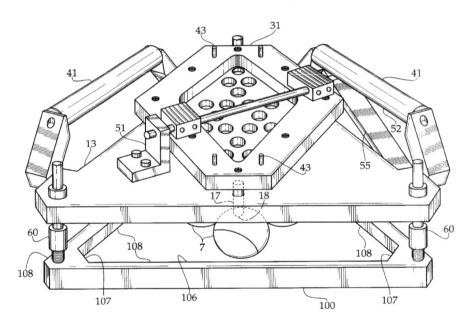

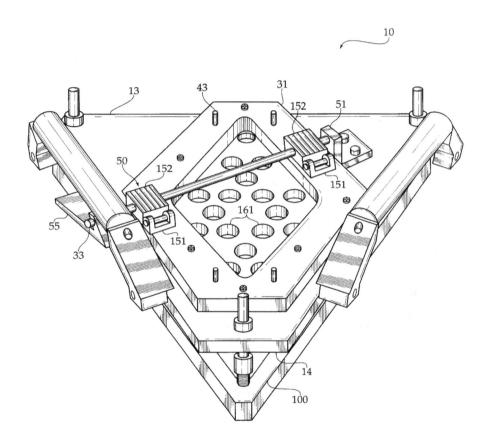

FIG.1

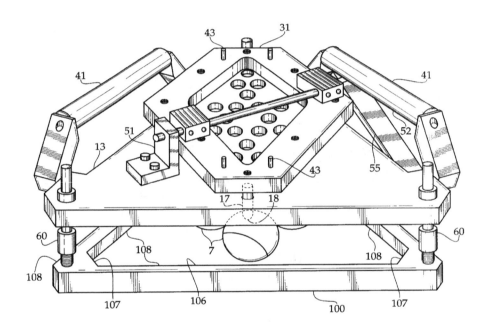

FIG. 2

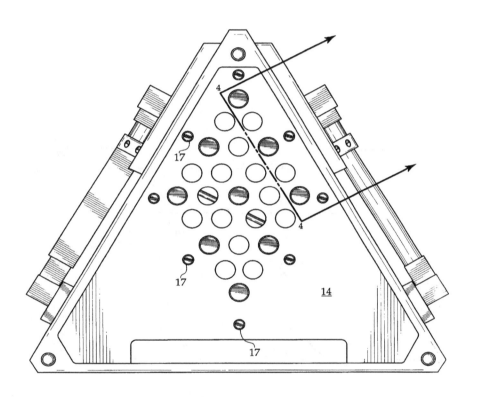

FIG. 3

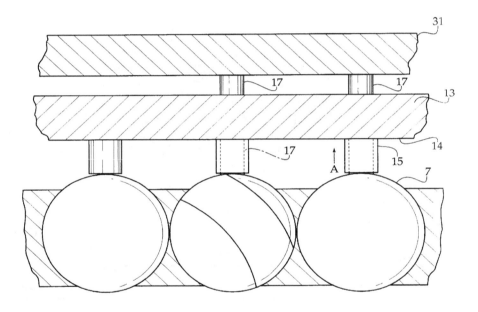

FIG. 4

5,997,404

1

**RACKING SYSTEM FOR ARRANGING
POOL BALLS**

BACKGROUND OF THE INVENTION

The invention relates to a system for racking pool balls on a pool table. More particularly, the invention relates to a racking system that allows players to tightly arrange pool balls in a compact formation on the pool table.

Pool games are extremely popular with people of all ages. Most games require that pool balls (often fifteen for playing "eight ball", but sometimes nine for the game of "nine ball") be arranged into a triangular or a diamond shaped pattern. The balls are arranged in a manner such that the apex of the formed diamond is located near a pre-marked spot on the pool table, which serves as a marker for placing the racked balls.

The pool balls must be arranged in a tight and compact formation to ensure that when it is first "broken", the balls do not deviate from their predictable lines of travel merely because of a poor racking of the pool balls. Unfortunately, there is no existing system that provides pool players with an easy mechanism for obtaining a tight formation of the desired shape on the pool table.

To rack the balls, most people use a triangular pool rack that holds the balls in its interior portion. The rack is removed once the balls are organized in a formation that roughly resembles the desired shape. Unfortunately, the size of the pool balls often lacks uniformity, which makes it difficult to properly rack the balls into a tight formation. As a result, the first player to break the rack of balls is at a disadvantage in that the improperly racked balls do not respond as would be anticipated from a properly racked set of pool balls.

Other devices have been proposed for the purpose of automatically racking pool balls or assisting therewith. Unfortunately, all of these devices operate by engaging the balls "en masse" in an attempt to press them together.

While these various devices attempt to provide simplified means for properly racking pool balls, none of these guarantee a tight and even formation. Additionally, these devices have a complicated structure and are expensive in construction, yet still fail to provide the desired results. Therefore, while these prior art units may be suitable for the particular purpose employed, or for general use, they would not be as suitable for the purposes of the present invention as disclosed hereinafter.

SUMMARY OF THE INVENTION

It is an object of the invention to provide a racking device for racking pool balls in a tight and compact manner on top of the pool table. Accordingly, each of the balls is individually engaged to urge them toward each other, and into a stable position on the pool table felt.

It is another object of the present invention to provide a simple mechanism for racking pool balls in a tight and even formation by providing a racking device that conforms to the shape and size of each individual pool ball. Accordingly, a racking device is disclosed that utilizes a plurality of sweeper pins to move the pool balls together into a tight formation, such that adjoining balls touch each other. The balls are then firmly pressed into the felt of the pool table by a plurality of pressure pins. Once the desired formation has been achieved, the racking device is removed without disturbing the racked pool balls.

To the accomplishment of the above and related objects the invention may be embodied in the form illustrated in the

2

accompanying drawings. Attention is called to the fact, however, that the drawings are illustrative only. Variations are contemplated as being part of the invention, limited only by the scope of the claims and their legal equivalents.

BRIEF DESCRIPTION OF THE DRAWINGS

The above and other aspects, features and advantages of the present invention will be more apparent from the following detailed description thereof, which is presented in conjunction with the following drawings, wherein corresponding reference characters indicate corresponding components throughout the drawing figures.

FIG. **1** is a diagrammatic perspective view of the racking device, viewed from the front.

FIG. **2** is a diagrammatic perspective view of the racking device, viewed from the rear.

FIG. **3** provides a bottom plan view of the racking device.

FIG. **4** provides a cross-sectional view taken in the direction of arrows **4—4** in FIG. **3**.

DETAILED DESCRIPTION OF THE PREFERRED EMBODIMENTS

FIGS. **1–4** show a racking device **10** that tightly arranges a plurality of pool balls on top of a pool table having a felt top surface.

According to the invention, the racking device **10** comprises a pool rack **100** that has an interior **106**, as shown in FIG. **2**. The pool rack **100** has three side walls **108**, wherein adjoining side walls **108** join together to form a corner **107**. The interior **106** of the pool rack **100** houses a plurality of pool balls **7** therein for arranging the balls **7** in the desired formation. According to the invention, the racking device **10** engages each individual pool ball **7** to obtain a tight and compact formation. While the present invention racks the balls **7** in a diamond shaped formation (comprising nine balls), the scope of the present invention is not limited by this aspect. Accordingly, it is possible to rack the balls in a triangular shaped formation (comprising fifteen balls) through minor adaptation of the present invention.

As shown in FIG. **4**, the racking device **10** comprises a pressure base **13** with a bottom surface **14**. A plurality of pressure pins **15** are mounted to the bottom surface **14**, extending vertically downward therefrom. The pressure pins **15** are spring mounted to ensure that they selectively move upward in the direction of arrow A when pressed on top of the pool balls **7**, but still exert pressure upon the pool balls **7**. The pressure pins **15** are positioned so that they each engage one of the pool balls **7** at the very top point thereof.

The racking device **10** has a pair of handles **41** that are mounted to and above the pressure base **13**, as shown in FIG. **2**. The handles **41** are substantially parallel to the pressure base **13**, such that when the handles are pushed downward, the pressure base **13** moves downward and causes the pressure pins **15** to press on top of the pool balls **7**.

The racking device **10** is provided with a sweeper arm frame **31** that lies parallel to the pressure base **13**. A plurality of sweeper frame guide pins **43** are mounted to the pressure base **13** and extend through the sweeper arm frame **31**. The sweeper frame guide pins **43** are mounted at approximately ninety degrees from the pressure base **13** to allow the sweeper arm frame **31** to move toward and away from the pressure base **13**, and to ensure that the sweeper arm frame **31** moves in a plane that is substantially parallel to pressure base **13**.

As shown in FIG. **4**, the sweeper arm frame **31** has a plurality of sweeper pins **17** extending vertically downward

5,997,404

| 3 | 4 |

therefrom. The sweeper pins **17** are spring mounted to ensure that the sweeper pins **17** can move upward when they contact the sides of the pool balls **7**. Each of the sweeper pins **17** has a conical tip **18**, as shown in FIG. **2**.

Each handle **41** has a bottom surface **52**. A sweeper lever **55** is pivotably mounted from the bottom surface **52** of one of the pair of handles **41**. As shown in FIG. **1**, the sweeper lever **55** is provided with an axle **33** that extends across the sweeper arm frame **31**.

A pair of sweeper activators **50** are mounted on top of the sweeper arm frame **31**. The sweeper activator **50** has a front portion **151** and a back portion **152**, wherein the axle **33** extends through the back portion **152** such that rotation by the axle **33** causes the back portion **152** to rotate in like direction. The front portion **151** is securely affixed to the sweeper arm frame **31**, while the back portion **152** normally lies in a plane that is substantially parallel to the sweeper arm frame **31**.

An axle mount **51** is provided on top of the pressure base **13** near the other handle **41**, which secures the axle **33** on the side opposite from the sweeper lever **55** pivotably mounted from the first handle **41**. Since the axle **33** is attached to the sweeper arm frame **31**, mounting the axle **33** to the axle mount **51** ensures that the sweeper arm frame **31** securely remains parallel to the pressure base **13**.

According to the invention, when the sweeper lever **55** is squeezed, the axle **33** rotates clockwise and raises the back portion **152**, which causes the front portion to pivot downward and move the sweeper arm frame **31** vertically downward. When the sweeper arm frame **31** moves downward, the sweeper pins **17** also move downward and each engages one of the pool balls off-center, pushing the sides of the pool balls **7**, causing the pool balls **7** to move inward against each other.

The sweeper lever **55** is spring mounted, such that when the sweeper lever **55** is released, it moves downward and away from the bottom surface **52** of the handle **41**, which causes the sweeper arm frame **31** to be raised upward and the sweeper pins **17** to release from contact with the pool balls **7**.

A plurality of springs are securely mounted between the sweeper arm frame **31** and the pressure base **13**. Thus, when the sweeper lever is released, the plurality of springs push the sweeper arm frame **31** upward and away from the pressure base **13**.

The racking device **10** has a plurality of guide assemblies **60** that secure the pressure base **13** to the side walls **108** of the pool rack **100**, as shown in FIG. **2**. It is preferable that the racking device **10** be provided with three guide assemblies **60** that are positioned at the corners **107** of the pool rack **100**. According to the invention, the guide assemblies **60** ensure that the pressure base **13** moves upward and downward toward and away from the rack **100** in a plane that is substantially parallel to the pool rack **100** and the pool balls **7** placed in the interior **106** of the pool rack **100**.

FIG. **3** provides a bottom view of the preferred embodiment of the racking device **10**, which arranges the balls in a diamond shaped formation for playing "nine ball." As is well known to most pool players, nine ball requires that nine pool balls **7** be arranged in a diamond shaped formation, wherein eight pool balls **7** are placed surrounding one pool ball **7** in the center thereof. Accordingly, the racking device **10** has eight sweeper pins **17** that extend downward through a circular hole provided in the pressure base **13**. Each sweeper pin **17** corresponds to the location of a pool ball **7** on the periphery of the diamond shaped formation. The racking device **10** is provided with nine pressure pins **15** that correspond to the location of each pool ball **7** in the diamond shaped formation, and are each positioned to engage one of the pool balls at its apex.

According to the invention, to obtain the desired pool ball formation, an appropriate number of pool balls **7** are placed in the interior **106** of the pool rack **100** on top of a pool table. The pool balls **7** are arranged in a configuration that roughly resembles the shape desired.

Next, the handles **41** are pushed towards the pool balls **7** on the pool table and the sweeper lever **55** is squeezed against the handle **41**. As the sweeper lever **55** moves upward, the sweeper arm frame **31** moves downward toward the pressure base **13**, and the sweeper pins **17** each engage one of the pool balls off center and press against the side of the pool ball **7**, which together pushes the pool balls **7** toward the center such that there is no gap between the pool balls **7**. As the balls move together, a tighter pool ball formation is obtained. Since the sweeper pins **17** are spring mounted to the sweeper arm frame **31**, the push force exerted by the sweeper pins **17** corresponds to the exact amount needed for moving the balls **7** into the tighter pool ball formation.

As shown in FIG. **1**, the pressure base **13** has a plurality of viewing holes **161**, which allows a use to view the pool balls **7** from the top as they are tightly arranged in the desired formation.

Once the balls **7** are tightly racked, the handles **41** are pushed downward for a few more seconds to press the pressure pins **15** on top of the pool balls **7** of the tightly arranged pool ball formation and push the balls **7** into the felt of the pool table. As the balls **7** are pressed into the felt, the sweeper lever **55** is released to allow the sweeper pins **17** to move upward and release contact with the pool balls **7**.

Once the user concludes that the pool balls **7** have been appropriately racked and firmly pressed into the felt of the table by looking through the viewing holes **161**, pressure on the handles **41** is released and the racking device **10** is removed upward from the pool table. The previous firm pressing of the pool balls **7** into the felt of the pool table ensures that the balls **7** do not move away from each other as the racking device **10** is removed therefrom. As shown in FIG. **2**, the racked pool balls **7** do not actually touch the side walls **108** of the pool rack **100**, which helps ensure that the removal of the racking device **10** does not disturb the pool ball formation.

Many specific details contained in the above description merely illustrate some preferred embodiments and should not be construed as a limitation on the scope of the invention. Accordingly, many other variations are possible within the spirit of the present invention, limited only by the scope of the appended claims.

What is claimed is:

1. A racking device for firmly arranging a plurality of pool balls in a desired formation on a pool table having a felt top surface, comprising:

    a pool rack having a plurality of side walls, wherein said side walls join together to form an interior;

    a pressure base mounted to the pool rack, said pressure base having a bottom surface;

    a plurality of pressure pins mounted on said bottom surface of said base, each of said pressure pins capable of firmly pushing one of the pool balls into the felt of the pool table;

    a pair of handles having a bottom surface, said handles mounted to the pressure base;

5,997,404

a sweeper arm frame mounted on top of said pressure base;

a sweeper attached to the sweeper arm frame lever mounted to the bottom surface of said handles for selectively moving the sweeper arm frame downward against the pressure base; and

a plurality of sweeper pins mounted from said sweeper arm frame, said sweeper pins extending downward such that the sweeper pins push the pool balls toward the adjoining balls when the sweeper arm frame moves downward toward the pressure base.

2. The racking device of claim **1**, further comprising:

an axle securely mounted from said sweeper lever, said axle extending across the sweeper arm frame; and

a pair of sweeper activators mounted on said sweeper arm frame, said sweeper activators securely attached to the axle extending across the sweeper arm frame.

3. The racking device of claim **1**, further comprising a plurality of guide assemblies, wherein the guide assemblies secure the pressure base to the side walls of said pool rack.

4. The racking device of claim **3**, further comprising an axle mount opposite said sweeper lever on top of the pressure base for securing the axle extending across the sweeper arm frame.

5. The racking device of claim **4**, wherein the sweeper activator comprises a front portion and a back portion, said front portion being securely affixed to the sweeper arm frame and said back portion lying on top of the sweeper arm frame.

6. The racking device of claim **5**, wherein the axle securely extends through the back portion of the sweeper activator, such that rotation by the axle causes the back portion to rotate in like direction.

7. The racking device of claim **5**, further comprising a plurality of sweeper frame guide pins that are mounted on the pressure base at a ninety degree angle, said sweeper frame guide pins extending through the sweeper arm frame for ensuring that the sweeper arm frame moves parallel to the pressure base.

8. The racking device of claim **7**, wherein the sweeper pins are spring mounted from said sweeper arm frame, such that said sweeper pins move upward when they contact the sides of said pool balls.

9. The racking device of claim **1**, wherein the sweeper pins are spring mounted from said sweeper arm frame, such that said sweeper pins move upward when they contact the sides of said pool balls.

10. The racking device of claim **1**, wherein the pressure pins are spring mounted from said pressure base for allowing the pressure pins to move upward when pressed on top of the pool balls.

11. The racking device of claim **3**, wherein the racking device is provided with three guide assemblies that are located at the corners of the pool rack.

12. The racking device of claim **11**, wherein the racking device is provided with nine pressure pins arranged in a diamond formation.

13. The racking device of claim **12**, wherein the racking device is provided with eight sweeper pins arranged in a diamond formation.

14. The racking device of claim **12**, wherein the pressure base is provided with a plurality of viewing holes for allowing one to see the pool ball formation from the top.

6

15. A racking device, for racking a plurality of pool balls, each having an apex, into a tight racked formation, upon a pool table having a felt top surface, comprising:

a pressure base;

a plurality of pressure pins mounted beneath the pressure base, the pressure pins positioned so that they each engage the apex of one of the balls when in the tight racked formation; and

a plurality of sweeper pins mounted to extend vertically beneath the pressure base, each of the sweeper pins is capable of vertical movement therethrough to selectively engage one and only one of the pool balls off-center from above and urge that pool ball inward so that the sweeper pins together press the balls tightly against each other.

16. The racking device as recited in claim **15**, wherein each sweeper pin has a conical tip.

17. The racking device as recited in claim **16**, further comprising a sweeper frame mounted for vertical relative movement with respect to the pressure base, a pair of handles mounted to the pressure base, and a lever for causing the sweeper frame to move toward the pressure base for moving the sweeper pins downward to engage the balls.

18. The racking device as recited in claim **17**, wherein the lever is located adjacent to one of the handles, so that the lever may be actuated by squeezing the lever toward that handle.

19. The racking device as recited in claim **18**, wherein the pressure base further has a plurality of viewing holes for allowing a user to see the position of the balls as they are being racked.

20. A method of tightly arranging a plurality of pool balls on a pool table having a felt top using a racking device comprising a pool rack, a pressure base having a bottom surface, a plurality of pressure pins extending downward from the bottom surface of the pressure base, a sweeper arm frame secured parallel to the pressure base, a pair of handles mounted on the pressure base, a sweeper lever attached to one of the pair of handles, an axle mounted from the sweeper lever that secures a pair of sweeper activators, the sweeper activators mounted on the sweeper arm frame, a plurality of sweeper pins mounted from said sweeper arm frame and extending downward through the pressure base, said method comprising the steps of:

(a) placing the appropriate number of pool balls in the interior portion of the pool rack;

(b) arranging the pool balls in a rough formation of the desired shape on the felt of the pool table;

(c) pushing the handles downward to push the pressure base downward for pressing the pressure pins on top of the pool balls;

(d) contacting the pool balls from above with the sweeper pins and pushing them inward such that there is no gap between adjoining balls;

(e) releasing contact of the sweeper pins from the sides of the pool balls; and

(f) terminating the push exerted on the pressure base for allowing the pressure pins to release contact of the pressure pins from the pool balls.

* * * * *

APPENDIX

D

Design Patent Example

US00D628386S

(12) **United States Design Patent**
Kohn

(10) **Patent No.:** **US D628,386 S**
(45) **Date of Patent:** ** **Dec. 7, 2010**

(54) **BUCKET LID TOOL ORGANIZER**

(76) Inventor: **Peter Kohn**, 3731 N. Country Club Dr., Apt 1223, Aventura, FL (US) 33180

(**) Term: **14 Years**

(21) Appl. No.: **29/349,057**

(22) Filed: **Mar. 4, 2010**

(51) **LOC (9) Cl.** **03-01**
(52) **U.S. Cl.** **D3/313**
(58) **Field of Classification Search** D3/304, D3/307–308, 310, 313, 315, 905; 220/500, 220/505, 904; D6/512–514; 206/557–566
See application file for complete search history.

(56) **References Cited**

U.S. PATENT DOCUMENTS

D417,785	S	*	12/1999	Daniels	D3/313
D426,921	S	*	6/2000	Renfrew	D28/73
D459,078	S	*	6/2002	Tondino	D3/313
D479,439	S	*	9/2003	Joss et al.	D7/553.4
D484,364	S	*	12/2003	Bailey et al.	D7/553.2
D508,634	S	*	8/2005	Chambers et al.	D7/553.8
D523,746	S	*	6/2006	Snedden et al.	D9/435
D541,531	S	*	5/2007	Leao	D3/308
D545,634	S	*	7/2007	Phythian	D7/637

* cited by examiner

Primary Examiner—Elizabeth A Albert
Assistant Examiner—Kelley A Donnelly
(74) *Attorney, Agent, or Firm*—Goldstein Law Offices P.C.

(57) **CLAIM**

The ornamental design for a bucket lid tool organizer, as shown and described.

DESCRIPTION

FIG 1 is a perspective view of the bucket lid tool organizer.

FIG. 2 is a right side elevational view of the bucket lid tool organizer.

FIG. 3 is a left side elevational view thereof.

FIG. 4 is a front elevational view thereof.

FIG. 5 is a rear elevational view thereof.

FIG. 6 is a top plan view thereof; and,

FIG. 7 is a bottom plan view thereof.

The broken lines are for the purpose of illustrating the environment of the design only and form no part of the claimed design.

1 Claim, 3 Drawing Sheets

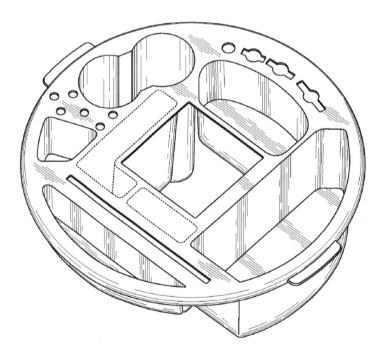

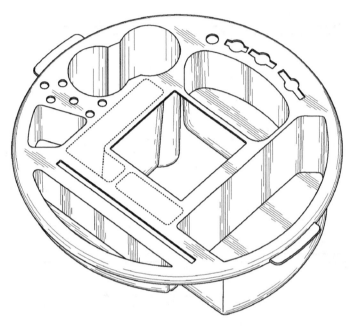

FIG. 1

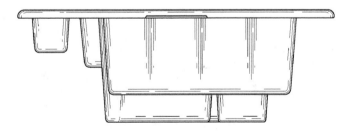

FIG. 2

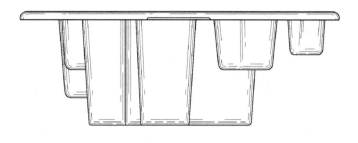

FIG. 3

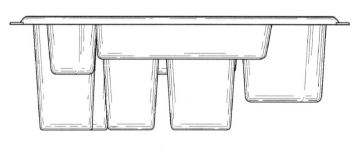

FIG. 4

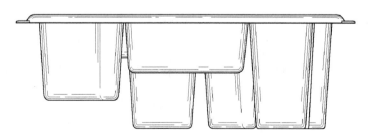

FIG. 5

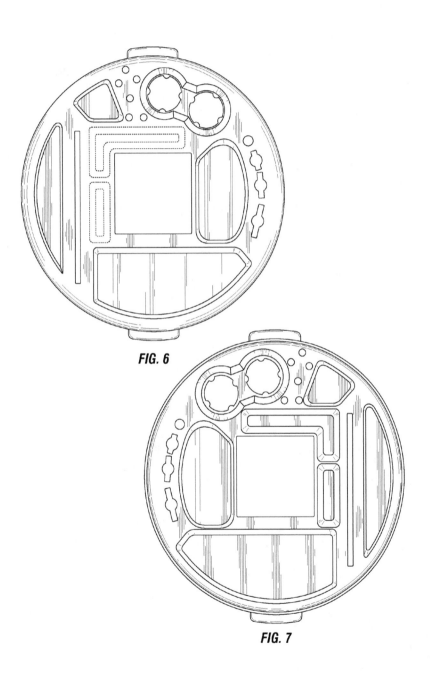

FIG. 6

FIG. 7

Glossary

"101 rejection"	rejection stating that an invention lacks patentable subject matter
"102 rejection"	rejection stating that an invention is not novel in relation to the prior art
"103 rejection"	rejection stating that an invention is obvious in light of the prior art
"112 rejection"	rejection due to issues with the manner or clarity of description contained in the specification and/or claims
abandoned	application that is no longer active is considered to be "abandoned," typically due to failure of the applicant to respond to an Office Action or notice that required response, or by the applicant specifically requesting that the Patent Office terminate proceedings regarding the patent application
Actual Use	basis for an application for a registered trademark, where you have used the mark in commerce before filing your application
alternative fee arrangements	generally used to describe modern and more progressive billing arrangements than traditional hourly billing
amendment	response that includes both changes and arguments/remarks (see "response")
America Invents Act (AIA)	patent legislation signed into law on September 16, 2011, which brought sweeping changes to US patent laws, including establishing the United States as a "first-to-file" patent jurisdiction
answer	response to a complaint, by a party who has been sued
apparatus claim	patent claim that focuses on the structural elements of an invention

appeal	application to a higher court, or an entity of greater authority, to reverse a decision made by a lower court or authority
application serial no.	(see "serial number")
argument	reason provided in response to overcome examiner's rejection
broad patent	covering more; potentially covering products sharing the same concepts (as in "broader protection")
business model	scheme, plan, or notion for how a profitable business will operate, including what activities are performed, what product is sold, what expenses are generated, and how a profit is created
cease and desist	request made to stop infringing activity
claims	see "patent claims"
class and subclass	category and subcategory employed to classify inventions; generally referring to the USPTO Manual of Classification, but may sometimes refer to other classification systems
classification search	research performed by identifying the possible category for an invention (generally as defined in the USPTO Manual of Classification), identifying all patents that have classified in that category, and then reviewing such patents for relevant prior art
common law protection	ability to prevent competitors from using a similar mark based upon the demonstrable recognition of your mark among relevant consumers
complaint	opening document that initiates a lawsuit by alleging the basis of the suit
composition of matter	combination of two or more substances, whether by chemical union or mechanical mixture
continuation-in-part	("CIP"); patent application that is filed after an initial application, which contains the same disclosure as the initial patent application, and also additional subject matter
copyright	exclusive legal right given to an originator (or entity assigned ownership by the originator) for a literary, artistic, or musical composition
criteria for patentability	requirements and standards upon which the decision to grant a patent is rightfully based
declaratory judgment action	("DJ action"); lawsuit filed for the purpose of having the court make a declaration, such as to declare that a patent is invalid or that no infringement is present
default judgment	judgment that may be taken against a party that fails to respond or meet a time deadline in a court case
design patent	protects the ornamental appearance of a useful object
discovery	process of exchanging information among the parties to a lawsuit, to collect facts and evidence
divisional	filed to pursue patent claims that were previously withdrawn because of a restriction (see "restriction")

election	document filed in response to a restriction, in which the applicant specifies which one of the inventions contained in a patent application will receive further examination
elevator speech	short, concise description of your invention, for provoking interest of others
enabling disclosure	description in a patent that provides sufficient detail to allow a person skilled in the art to carry out the claimed invention without undue experimentation
equitable estoppel	bar to a party from asserting a claim, in the context of patent litigation, generally because of statements made that the other party relied upon
European Patent Office	("EPO"); central patent office having mechanisms for establishing patent rights for all member countries within Europe
examiner's amendment	changes made in the patent application by the examiner, typically with approval of the inventor's representative
examining groups	sections within the Patent Office that are designated for reviewing applications that describe a certain type of technology
extension fee	fee for extending the period of response, when permitted, if a response is required by the USPTO
filing date	date that a patent application is considered filed; may differ from the date that an application was first submitted if the application was incomplete initially
Final Office Action	Office Action issued after the inventor has responded to a non-final Office Action (see "Office Action")
first-to-file system	any patent system that prioritizes the first entity to file a patent application
fixed fee	billing arrangement whereby the amount paid for designated work remains fixed, regardless of the amount of time it actually takes to do the work
foreign filing license	included on a patent application filing receipt; provides permission to file a patent application in foreign countries
functionality	how well a product provides/fulfills its intended function
hourly billing	the most typical client billing system, where a client pays only for time actually spent by the attorney and staff, as well as disbursements for expenses incurred
inequitable conduct	when a patent applicant has made a "material misstatement or omission with intent to deceive"; may be the grounds for invalidating a previously granted patent
infringement	violation of the patent rights that another person or entity holds
infringing	violating the patent rights that another person or entity holds
initial examination	administrative reviewed to determine if all necessary documents are included, whether the requirements for a filing date are present, and whether the patent application is otherwise complete

injunction	court order that prevents a person or entity from continuing or starting a prohibited action
intellectual property	(IP); intangible property, generally an aspect of creativity; the field of the law that deals with the protection of ideas and similar intangible property; patent, trademark, copyright, and trade secret are generally categories of intellectual property
Intent-to-Use	basis for an application for a registered trademark, where you have not yet used the mark in commerce before filing your application but assert a genuine intent to begin doing so
International Trade Commission	(ITC); US federal agency that has the jurisdiction to make determinations regarding unfair importation of products; provides a forum for determining when products that infringe a US patent can be blocked from importation by US Customs
inventorship	determination of who are the correct inventors to be named in a patent application filing
issuance fees	fees that must be paid after a Notice of Allowance for the patent to be issued
large entity	for the purposes of paying USPTO fees, an organization with 500 or more employees
licensee	person or entity that licenses patent rights from the patent holder (see "licensing")
licensing	providing permission for another to use something owned
litigation	legal proceedings
machine	invention with parts that interact with each other
maintenance fees	fees required to keep a patent in force; required to be paid 3 1/2, 7 1/2, and 11 1/2 years after patent issuance
manufacturability	feasibility of manufacturing a product using known manufacturing techniques
manufacture	physical structure that is so configured for a functional purpose
marketability	ability of a product or service to match the needs and desires of the relevant market
method	series of steps for achieving a result (see "process")
method claim	patent claim that focuses on the steps in producing a result or carrying out a process
micro entity	for the purposes of paying USPTO fees, must qualify as a small entity and have filed fewer than four (4) patent applications previously (not including provisionals) and have an income less than the threshold level ($160,971 in 2016)
monetize	convert into money or make money from
motion to dismiss	request made to a court or other tribunal to terminate a lawsuit or similar action
narrow patent	covering less; likely only covering products sharing the same details (as in "narrower protection")

non-obvious	sufficiently inventive or unexpected such that it would not be obvious to a person of ordinary skill in the field of the invention
non-publication request	made when filing a utility patent application, requesting the Patent Office to refrain from its typical practice of publishing the application eighteen months after its earliest priority date
Notice of Allowance	notice indicating that a patent application has been approved
notice of missing parts	generated after initial examination, if the application is found to be incomplete
novel	new; in terms of patents, different at all from the prior art
objection	reference to a technical violation found during review of a patent application; typically made for minor deviations from the Patent Office rules and patent laws
obvious	anyone practicing in the field of the invention would know that something could be done in a particular way (see "non-obvious")
Office Action	official communication from the Patent Office, detailing the examiner's review of the patent application
oversearch	made-up term to describe searching beyond any reasonable benefit
patent	legal protection that gives the patent holder the right to prevent others from making, using, or selling the invention or innovation covered by the patent for a specified period of time
patent application	document filed in the Patent Office to officially request a patent
patent application publications	patent applications that have been published
patent claims	portion of a patent that sets forth the scope of protection and defines the invention for purposes of determining when infringement would be present (see "infringement")
Patent Cooperation Treaty	(PCT); provides mechanisms and procedures for preserving patent rights in designated foreign jurisdictions, to allow for the future filing of patent applications in those jurisdictions
patent drawings	illustrations contained in a patent application that depict the invention, and sometimes its environment
patent examiner	employee of the Patent Office who reviews patent applications and any arguments for their approval
patent infringement lawsuit	brought by a patent owner against a person or entity allegedly violating a patent
patent invalidity	finding that a patent was issued in error, and thus is not valid
patent issuance	official granting of a patent
patent litigator	attorney who represents clients in patent lawsuits and related transactions
patent no. (patent number)	number assigned to a patent by the issuing patent office; not the same as the serial number that is assigned to the patent application (see "serial number")
Patent Office	United States Patent and Trademark Office (USPTO)

patent pending	status afforded an invention when a patent application has been filed and is not yet abandoned or issued as a patent
patent portfolio	collection of patents owned by an entity
patent process	legal process for obtaining a patent
patent rights	rights provided to a patent holder by a patent, which can be transferred temporarily to a licensee or permanently via sale of the patent (see "patent," "licensee")
patent search	process of researching existing patents; also called a "prior art search" (see "prior art")
patentability	ability of an invention to qualify for patent protection
patentable subject matter	proper type of invention on which a patent may be based
Petition to Make Special	request to the Patent Office to have the application granted "special status" so that it will be examined sooner than would be expected otherwise
photo-realistic renderings	computer-generated drawings depicting an item and appearing to the eye as if they were photographs of the physical item
poor man's patent	age-old myth that the postmark on an invention description, mailed by the inventor to him- or herself, will serve as proof of invention; patently untrue
post-grant review	category of various processes used to have the validity of a patent reconsidered after its issuance
prematurely expired patent	patent that has expired before the end of its natural term, generally due to the non-payment of maintenance fees
prior art	information describing relevant inventions that came before the patent application currently being considered
prior art references	documents that are considered prior art, such as patents and publications
prior art rejection	rejection based upon a comparison of the claimed invention to the prior art; typically communicated in an Office Action
priority date	assigned by the Patent Office; used to determine what is considered prior art and who is considered the first to file (see "prior art"; "first-to-file system")
process	series of steps for achieving a result (see "method")
proprietary	owned or capable of ownership
prototype	first, or preliminary, model of something
provisional patent application	filed to establish priority for an inventor at the USPTO; a utility patent application filed within one year of the provisional can claim priority from the date of the provisional application
public disclosure	revealing information about an invention to the public, or publicly displaying or selling, an invention
public domain	belonging to the public as a whole; free to use by anyone

reexamination	procedure by which a granted patent is again examined to make a new determination of patentability (see "patentability")
registered trademark	trademark that has been registered by the USPTO
response	document filed to respond to an Office Action; typically required within a specified time frame
restriction	determination by an examiner that a patent applications contains more than one invention, and requiring that the applicant choose only one for further examination
petition to revive	request filed with the USPTO, along with the appropriate fee, to re-instate an abandoned patent application or prematurely expired patent, generally by asserting that the abandonment or delay in paying the maintenance fee was unintentional or unavoidable
royalty	fee paid by licensee for temporary grant of patent rights (see "patent rights"; "licensee")
scope	extent of protection (as in "scope of protection")
serial number	unique designation assigned to each patent application when it is filed
small entity	for the purposes of paying USPTO fees, an organization with fewer than 500 employees
specification	written portion of the patent application, excluding the claims; also known as the "spec" or the "disclosure"
statement of use	(or "use in commerce"); declaration of bona fide use of a mark in the ordinary course of trade, and not made merely to reserve a right in a mark
substantive examination	detailed review of a patent application on the merits of the invention as described (compare to "initial examination")
summary judgment	judgment entered against another party without a full trial
Track One	program that allows a patent applicant to pay a surcharge to speed up the examination process
trade secret	information that provides an economic advantage to a business and is maintained in secrecy by the business
trademark	(or "mark"); aspect of a product or service that distinguishes it in the marketplace
United States Patent and Trademark Office	(USPTO, aka "the Patent Office")
USPTO	United States Patent and Trademark Office ("the Patent Office")
utility patent	protects a functional invention or a functional improvement to an existing invention
virtual prototypes	computer-generated visual representation of an item that depicts the operation of the item as if it were physically constructed

About the Author

Richard W. Goldstein is a registered patent attorney and founder of Goldstein Patent Law. Over the past 20+ years, he has represented thousands of entrepreneurs from across the United States, and helped them obtain nearly 2,000 patents. He has educated tens of thousands more about the patent process. In addition to writing about patents, he frequently writes and speaks on topics related to leadership, collaboration, sales, and marketing. His personal blog is: www.richgoldstein.com and his firm website is www.goldsteinpatentlaw.com.

About the Illustrator

Thom Wright is a patent illustrator, designer, and prototype builder. Examples of his work can be found at: www.inventionart.net.

Index